Insights into Mathematical Thought

Excursions with Distributivity

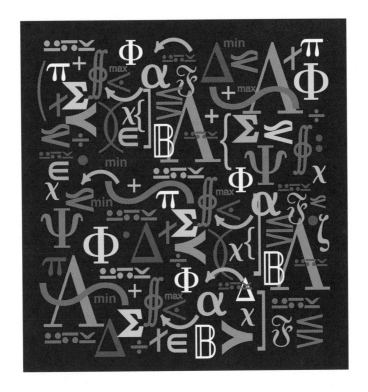

Stephen I. Brown
Professor Emeritus
University at Buffalo, State University of New York
Buffalo, New York

NATIONAL COUNCIL OF
TEACHERS OF MATHEMATICS

Library of Congress Cataloging-in-Publication Data

Brown, Stephen I.
    Insights into mathematical thought : excursions with distributivity / Stephen I. Brown.
    pages cm
    Includes bibliographical references.
    ISBN 978-0-87353-712-4
1. Distributive law (Mathematics). 2. Algebra, Abstract. I. Title.
    QA162.B76 2012
    515′.782—dc23

                            2012021403

The National Council of Teachers of Mathematics is the public voice of mathematics education, supporting teachers to ensure equitable mathematics learning of the highest quality for all students through vision, leadership, professional development, and research.

Printed in the United States of America

DEDICATION

With unbound gratitude and deep sadness, I dedicate this book to the memory of my dear friend of forty years, Joseph Sander Lukinsky. We met as graduate students, and it did not take long to appreciate that even though he had been enmeshed in studying ethics, and I mathematics, we shared a deep interest in education writ large. We encouraged each other to seek alternatives to taken-for-granted assumptions about teaching, learning, and subject content.

Although we each had essentially no prior experience in investigating the connections between our diverse fields—fields that most of our friends and colleagues thought of as incompatible—we began teaching courses together. They went under variations of the title Math and Morality. Muddling our way through said courses, we adopted an ever so slight defensive air when colleagues asked, "What exactly did you say you were planning to teach?"

In addition to the mind-expanding experience of joining two disparate fields, our relationship had another aspect that I came to cherish. Joe, who was a hair's breadth from becoming a professional baseball player (opting for the rabbinate instead), decided that it was time for me to learn to think of a ball as more than a flexible spheroid. He taught me to play racquetball—no mean feat considering that, as a kid, I was always the last selected to play stickball in the streets of Brooklyn.

This body-expanding experience morphed into a mind-expanding one, apparent to anyone who viewed us from the gym track above our racquetball court. We often interrupted our games to continue an unfinished conversation from the day before. More often than not, we were unaware that we were engaging in a bizarre sense of playing racquetball. We were reminded of this pattern however, by the observers "from above"—literally speaking, of course.

We had created a new sport, perhaps dubbed a math/moral racket. In addition to having a ball, we eventually turned our conversations to explore not only what we were creating between two ostensibly different fields but also the power of doing so in the context of sports. Imagine how much less aggressive football would be if players regularly interrupted it, stopping the clock just to discuss what they were learning about the game (and themselves) without being devastated by its jeopardizing their score. Of course it sounds absurd, but can you transform that into something analogous that worked for our racquetball game?

I can imagine how much more enticing this book would have been had my friend lived long enough to experience the labor pains of its publication.

CONTENTS

PREFACE

To see a world in a grain of sand
And a heaven in a wild flower;
Hold infinity in the palm of your hand,
And eternity in an hour.

—William Blake

Central Theme

A grain of sand: an apt metaphor for this book. It is a grain that has invaded my own psyche in ways that have surprised me over the years. That grain is the distributive property:

$$a \bullet (b + c) = (a \bullet b) + (a \bullet c)$$

It is a property of arithmetic that some take for granted but is sometimes a challenge for beginners in mathematical thinking to understand. Regardless of the level of sophistication, however, it is rarely appreciated in a generative sense against the larger mathematical and educational terrain.

Is basing an entire book primarily on one property a bit excessive? Perhaps so, but if we keep in mind that the property is a peephole on the world of teaching, learning, and mathematical thinking, then it should have some surprising generativity. In fact, it is invested with the potential to generalize in the sense that studying one person (one's child, for example) from many different perspectives enables us to understand important things about humanity, even though all people are different. It is not so much that by studying one person well we will identify elements common to all people; rather, we reveal categories of thinking about that person that may enable us to better understand others.

Though the distinction may be a bit exaggerated, I have sought mathematical categories that focus less on technical matters and more on ones that portray the nature of mind and creative thought. Though focusing on the distributive property as foreground, I will offer two themes as an undertow of sorts: problem posing and mathematics as a humanistic enterprise. These categories are clothed in many different forms in the bibliography below.

Overview

Although the distributive property is the centerpiece, this book not only covers a variety of content but also encompasses many different orientations and levels of abstraction and intensity.

I did not set out to write a book with this property in mind. In fact, the distributive property selected me before I selected it. Over the years I had written about several issues and problems in education in general and in mathematics education in particular, and only in a few of them had I focused intentionally on the distributive property. Rather recently, I realized how much the property had snuck its way into what I was thinking and writing about. To be sure, some pieces that I had written focused head on with the grain of sand. I have revised and incorporated some of them into this collection, as indicated in some of the references, but I did so to broaden their scope and to integrate them with some of the themes of this book. I believe you will profit from the evolution of my own thinking about the relationship.

The book begins gently, showing how many people who are not formally familiar with the property are in fact already aware of it and use it implicitly in much of their arithmetic calculation. Accompanying that realization is a collection of several different models (some concrete, others more abstract) that offer an intuitive understanding of the more formal version of the property.

We move from those excursions in chapter 1 to a more playful orientation in the second chapter. By intertwining geometric and algebraic renditions of the distributive property, I will relate it to topics ranging from prime numbers to mental arithmetic and number tricks—some of which may be new to you.

Chapter 3 explores how a small creative change can lead to something unexpected. By paying attention to a simple set of observations and by seeing those observations in a new light, we end up with a new algorithm for multiplying numbers: one in which the distributive property plays a prominent role.

Chapter 4 challenges our intuition. Although the early focus of the book was to entice you while honoring your intuitions, this chapter is designed to place you in a position of finding discomfort with a tightly held intuition. What do we do (and the answer may be different for all of us) when our intuitions go awry?

The next three chapters use the distributive property a bit more abstractly. We begin not only to use the property but also to see the property itself as a source of inquiry. In chapter 5, we view the property as a form of mapping rather than merely as a statement.

The property is viewed as a stepping stone to get at the concept of similarity and difference: a fundamental idea that pervades our thinking in many mathematical and non-mathematical domains.

Chapter 6 further highlights the property's role in an important structure within which it is often embedded: a field. The distributive property is different in an important way from all the other properties of a field, and that makes it a prime candidate for raising abstract questions about its role in the structure it is embedded in.

In chapter 7, we take a panoramic view of several different mathematical areas and investigate how the distributive property operates within them. We begin with a description of the extension to imaginary numbers and point out that the concept of extension is more problematic than many have come to believe. For the integers, for example, we see how that property influences the rules established for multiplying negative integers. We compare its role with what at first appears to be a more intuitively appealing approach. The two approaches have more in common than we would have predicted.

Chapter 7 highlights aspects of mathematics that are more personal than we are used to associating with the discipline. The chapter leads us to wonder and hypothesize about how some fascinating ideas in number theory evolved, and we can find humor in exploration normally viewed as focused exclusively on logic.

The last section of chapter 7 takes a reverse orientation from that of the first two sections. Instead of focusing on aspects of nineteenth- and twentieth-century expansions of earlier conceptions of number properties, it takes a retrospective worldview of mathematical thinking that is more than two millennia old: Euclid's concept of area and length. The Pythagorean theorem is seen though Euclid's eyes, and it takes a while to fully appreciate how he operated without such modern-day trappings as the formula for the area of a square or of a triangle. One might think that he was limited in what he could convey about area. The situation was quite the contrary, and anyone who thinks that the Pythagorean theorem is best captured by an algebraic statement ($c^2 = a^2 + b^2$) or even by a statement about squares is in for a surprise. Euclid made generous use of various extensions of the distributive property from a geometric point of view.

Finally in chapter 8, we attempt to shed light not only on the distributive property but also on the style and format of this book itself. It is a helpful heuristic to do so by comparing this book with the Common Core State Standards for Mathematics (CCSSM)—a document in mathematics education inspired by an effort on the part of states to coordinate K–12 curriculum on a national level. The CCSSM project is significantly more ambitious than this book, and though chapter 8 does not evaluate CCSSM, it will elaborate on themes that might enrich the enterprise. Though focusing on subject matter content and student practice, CCSSM is open regarding teaching strategies.

Audience

This book has several intended audiences. They range from teachers of mathematics at several different levels (mainly secondary school but including parts of interest to those teaching at lower and higher levels) to interested citizens wanting a glimpse of the nature of mathematical thinking. I invite readers at both extremes to view aspects of mathematical thinking that combine the search for logical dimensions with aesthetic and interpersonal ones as well.

Because we have cast a wide net by using the distributive property as a peephole, teachers may be able to incorporate some of the content of this book within topics that already exist in the curriculum. For example, different algebraic variations of the property are intertwined with geometric formulations of the concept. We have used that intertwine in the context of presenting enticing problems, some of which are thereby accessible to younger students.

Though prime numbers are part of the curriculum (such as their infinitude, unique factorization of composites into a product of primes, and the ability to reduce all fractions to lowest terms), by focusing on the distributive property within the natural numbers, we appreciate anew what we had taken for granted in the natural numbers—something that might interest teachers at various levels.

Beyond specific content, teachers at all levels may profit from strategies to incorporate aspects of thinking and feeling that mathematical activities rarely include. As mentioned, we will connect mathematics and humor. With just a little digging, we can encourage students to discuss what they find humorous—ranging from peculiar to outright laughable. This is not to belittle the subject matter but rather to indicate that it is part of the human scene.

For most subject matter, one way to expose what is both fascinating and humorous about it is to get a glimpse of its evolution. When we present subject matter as if it were born full blown like Athena from Zeus's head, with no labor pains, we not only misrepresent the evolution of ideas but also convey to students that misunderstandings are a particular weakness of theirs. In several places, this book points out that the language of number systems suggests that something took place historically that is not quite so polished as it appears in most texts. An interesting discussion in almost any section that uses the extension of numbers could begin by asking students why such numbers are labeled "real," "complex," "imaginary," "irrational", "negative," "transcendental," and so forth. Although a great deal of historical writing addresses issues involving extension of number systems, one need not be an expert to make observations about the language and to encourage people to wonder what it might signify.

To make progress in fields that resisted movement, mathematicians needed not only the passage of time but also the courage to notice, ask about, and pursue unpopular ideas.

Many topics this book explores use "what if not" thinking. Without making breakthroughs in redefining fields, we indicate how even minor tweaking of what is taken for granted can lead to highly creative activities and new understandings. As mentioned, this book offers the example of seeing a pattern in a new light as the inspiration to develop a new algorithm for multiplying positive integers.

Those interested primarily in elementary education might find much of chapters 1 and 2 an appropriate start. Depending on what captures your imagination in the initial reading, chapter 3 and parts of chapter 7 ("Multiplying the Integers") might be worth exploring as well. The introduction to chapter 5 offers a glimpse at the relationship of mathematical thinking to everyday life and might inspire them to embellish much of what they teach with ways of incorporating the concept of "same/different" in their teaching. Chapter 8 offers a good overview and summary of the book, and you might enjoy delving into parts therein of particular interest.

People most interested in education at higher levels might begin with chapter 1 (which offers many ways, probably new, to view the distributive property). Then chapters 4, 5, and 6 in any order might inspire innovative ways of viewing aspects of the curriculum. The section in chapter 7 on primes in different domains would add a new perspective on much of what is taken for granted about primes. The discussion of Euclid's proof of the Pythagorean theorem in relation to a modern point of view may afford a new appreciation for Euclid's genius—one appreciated in the absence of using real numbers to make sense of area.

You need not read this book sequentially to appreciate it. Though chapters occasionally refer to each other, they are mostly self-contained. Feel free to read this book by intertwining skimming with careful reading. Though I use language and symbolism that may sometimes be unfamiliar, I have tried to buttress such content with analogies, metaphors, and sketches to foster an intuitive feeling for seemingly complicated and abstract concepts. If the mathematics looks too technical to handle at times, you may want to skip the details and read about the significance of the ideas of the chapter before deciding how much time to invest in the earlier sections. Just as teachers encourage their students to work cooperatively, they might profit from reading this book with colleagues. If you are a teacher and can find someone teaching at a different level, consider sharing what you find enlightening, threatening, irrelevant, impractical, ennobling, and so forth. Readers intrigued by the format and content of this book but who are not professional educators might enjoy discussing various sections with people who have diverse backgrounds and interests in the sciences and the humanities.

Permissions

I have received permission to reproduce several diagrams from other works of mine:

- Modifications of figures 3.1–3.3 and 3.7, which appeared in "A New Multiplication Algorithm: On the Complexity of Simplicity," *Arithmetic Teacher* 22 (November 1975): 546–54.

- Table 5.2 appeared in *Reconstructing School Mathematics: Problems with Problems and the Real World*, Peter Lang Publishing Inc., New York, 2001. Figures 5.1 and 5.2 first appeared in "Transcending the Kpelle Nightmare: Personal Evolution and Excavations," *Philosophy of Mathematics Education Journal* 22, November 2007 (available from http://people.exeter.ac.uk/PErnest/pome22/).

- Figures 6.1 and 6.3 appeared in "Multiplication, Addition, and Duality," *Mathematics Teacher* 59 (October 1966): 543–50, 591.

Bibliography

Brown, Stephen I. "Mathematics and Humanistic Themes: Sum Considerations." *Educational Theory* 23 (Summer 1973): 191–214.

———. "The Logic of Problem Generation: From Morality and Solving to De-posing and Rebellion." In *Girls into Maths Go,* edited by Leone Burton, pp. 196–222. Sussex, England: Holt, Rinehart, and Winston, 1986.

———. "Humor in E(arnest): Relatively and Philosophically E(a)rnest." In *Festschrift in Honor of Paul Ernest's 65th Birthday,* edited by Bharath Sriraman, pp. 97–128, The Montana Mathematics Enthusiast: Monograph Series in Mathematics Education, University of Montana. Charlotte, N.C.: Information Age Publishing, 2009.

Brown, Stephen I., and Marion I. Walter. *The Art of Problem Posing.* Philadelphia: Franklin Institute Press, 2005.

National Governors Association Center for Best Practices (NGA Center) and Council of Chief State School Officers (CCSSO). *Common Core State Standards for Mathematics. Common Core State Standards (College- and Career-Readiness Standards and K–12 Standards in English Language Arts and Math).* Washington, D.C.: NGA Center and CCSSO, 2010. http://www .corestandards.org/.

ACKNOWLEDGMENTS

I am grateful to several people and organizations that have been supportive and gently critical at various stages that directly and indirectly affected this project: students, colleagues, family members, editors, and funding agencies.

Joanne Hodges (director of publications) and Myrna Jacobs (publications manager) at the National Council of Teachers of Mathematics offered helpful editorial advice. Albert Goetz (editor of *Mathematics Teacher*) made helpful comments on the style and substance of an earlier draft. Gabe Waggoner was my editor for this book, and I have not received a more sympathetic, empathetic, careful reading from any book editor in my previous publications. I am also most grateful to Randy White (production manager) for graciously making changes to the text at the eleventh hour.

I have especially enjoyed picking up subtle (sometimes nonverbal) reactions from my own offspring, Jordan and Sharon, who were the first guinea pigs to experience my unconventional teaching style (and who are now learning to cope with such matters with their own kids). Anything from a wrinkled nose to an upset stomach to a smile or even a guffaw (as when they, aged ten and twelve, observed me teaching a class on the belly button problem) have influenced my inclination to incorporate, revise, or toss out some ways of thinking and teaching that appear in one form or another in this book.

I am most grateful to Larry Copes and Fran Rosamond, former students, who have held a more recent mirror to my thinking by producing a book (*Educational Transformations: The Influences of Stephen I. Brown*) featuring more than forty essays by friends, family, colleagues, and former students who reacted to my teaching and writing. Critical comments regarding references to both substance and style of my teaching and writing influenced me while writing this book.

In addition to the many students and colleagues who contributed significantly to my personal and intellectual growth in that collection, others include Bob Davis, Mary Finn, Norton Levy, Edwin Moïse, David Nyberg, Israel Scheffler, and David Wheeler. Harry Lucas Jr. and Albert Lewis have been most helpful in updating me regarding R. L. Moore's continuing legacy.

Tom Giambrone, a former student, now a professor himself, has been enormously helpful in encouraging me to enter the twenty-first century from both a conceptual and a practical point of view with regard to the potential and misuse of technology. He has come to understand how to make use of and overcome the inner workings of computers, electronic printers, and the like. Moreover, in an overly patient way, he has led me to believe that these devices act as if they had not only brains but also minds.

Finally, I am eternally grateful to three people: (1) Eileen Mae Thaler, my high school sweetheart, who knew how to criticize and encourage me from the time we met as teenagers in 1116—Miss Titcomb's homeroom class—as freshmen at Erasmus Hall High School in Brooklyn; (2) Mrs. Eileen T. Brown, the teacher extraordinaire, in my middle years whose brilliant criticism and passionate encouragement were without limits; and (3) Dr. Eileen Thaler Brown, the psychologist who in my dotage has learned how to make it all soar by attending in the most passionate and subtle ways to the differences between her J and my P (Myers–Briggs test).

CHAPTER 1

The Distributive Property: Multiple Ways of Seeing

Mathematics and Personality

Many people believe that though mathematics may be challenging and interesting (or even threatening), it lacks personality. Those inclined to think in images, metaphors, and other devices normally associated with poetry know, however, that even numbers (and, of course, odds) can exhibit an anthropomorphic sort of behavior. Just as the deaf hero in André Gide's *La Symphonie Pastorale* "hears" music by imagining how colors would sound, a good friend of mine associates natural numbers with different nuances of color and form. For example, she remembers birthdays by associating color with them.

Several years ago, *Arithmetic Teacher* published articles that tried to describe the unique personality of numbers. "What You Always Wanted to Know about Six but Have Been Afraid to Ask" (Hoffer 1973) spoke of properties unique to the number 6 (for example, it is the smallest "perfect number") as well as some that all numbers share. The article ends with the following delightful paragraph: "As six education becomes a richer subject in the schools . . . we shall be able to liven up many parties and conversations with the notion of six. It is time now, in our society, that we hear resoundingly the completion of the immortal saying, 'Two is company, three is a crowd, and six is a riot'" (p. 180).

Hohfeld and Schwandt (1975) further elaborated on the cult of six. Not to be outdone by "six education," Avital (1978) published "The Plight and Might of the Number Seven." Toward the beginning, Seven comments, "I am not *perfect*, but I am *prime* and Six isn't" (p. 22). Avital then pursues several other ways in which Seven can outperform Six. To start, he points out that Seven can be represented as a difference of two squares ($4^2 - 3^2$), whereas Six cannot.

A more recent elaboration is *7: The Magical, Amazing, and Popular Number Seven* (Eastis 2011). Though offering relatively few explanations, the book offers many pictures and some historical and pop-culture documents that depict in great splendor both mathematical and nonmathematical appearances of the number 7. Eastis invites the reader to guess the saliency of each portrayal involving the number 7, even including an answer sheet at the end.

Another contribution to mathematical personality comes from a wonderful story told by G. H. Hardy (1877–1947), a leading British mathematician, about Srinivasa Ramanujan, an Indian mathematician who made groundbreaking but untutored mathematical observations, often using unconventional symbolism. In *The World of Mathematics*, Newman (1956) quotes Hardy as saying about Ramanujan, "I remember once going to see him when he was lying ill at Putney. I had ridden in taxi-cab number 1729, and remarked that the number seemed to me rather a dull one, and that I hoped it was not an unfavorable omen. 'No,' he replied, 'it is a very interesting number; it is the smallest number expressible as a sum of two cubes in two different ways'" (p. 375).

Just as we can find personality in mathematics by focusing on how objects (such as numbers) behave, we can relate specific behavior to objects and to other behaviors. (*Behavior* refers here to the mathematical properties of a set of objects and operations.)

One such behavior has enchanted me for several years, illuminating many aspects of my mathematical thinking: the distributive property. It is only one of many numeric properties explicitly taught to, used to entice, or dangled before students from grade school through graduate-level mathematics. Nevertheless, it features prominently in several different ways, only some of which the literature and the curriculum on mathematical systems have discussed.

One can view the distributive property in many different systems. But let's recall how it functions in its simplest form—often disguised—in a number system familiar to students as early as third or fourth grade in their study of the counting numbers. The distributive property tells us the following with regard to addition and multiplication:

For all natural numbers a, b, and c, $a \cdot (b + c) = (a \cdot b) + (a \cdot c)$.

Invoking *Le Bourgeois Gentilhomme*

We have many ways to look at what that property asserts and how to interpret and apply it. Imagine that you are in the audience of *Le Bourgeois Gentilhomme* (*The Middle-Class Gentleman*), a Molière comedy–ballet first presented to the French king in 1670.

In the play, Monsieur Jourdain is a pretentious middle-class social climber who would like to rise above his station to become part of the aristocracy. Among other things, he hires a philosopher to enable him to express himself as a member of the upper class. Following is a loose English translation of an interchange between them (Jones 2008).

Jourdain: I'm in love with a lady of great quality, and I wish that you

	would help me write something to her in a little note that I will let fall at her feet.
Philosophy Master:	Very well. Is it verse that you wish to write her?
Jourdain:	No, no. No verse.
Philosophy Master:	Do you want only prose?
Jourdain:	No, I don't want either prose or verse.
Philosophy Master:	It must be one or the other.
Jourdain:	Why?
Philosophy Master:	Because, sir, there is no other way to express oneself than with prose or verse.
Jourdain:	What! When I say, "Nicole, bring me my slippers, and give me my nightcap," that's prose?
Philosophy Master:	Yes, sir.
Jourdain:	By my faith! For more than forty years I have been speaking prose without knowing anything about it, and I am much obliged to you for having taught me that.

Now for the distributive property in Monsieur Jourdain fashion. If you have not come across the principle explicitly before, assume that you are attending a performance in French, and though you have some knowledge of the language, it is harder to follow than you expected. You are about to doze off when you notice that all the seats in the theatre are occupied, and you wonder whether the other attendees too are feeling a bit disengaged. You are sitting in the 29th seat in row 21 of the theatre. Four rows are behind yours. You entered at seat number 1 in your row and notice that just eight more seats are beyond yours (number 29) to the end of your row. You assume that all the rows before yours have been listed in order from row 1 and that each row has the same number of seats. So you know how many rows there are and how many people are in each row. As your last official conscious act, you decide to figure out how many people are in the theater (awake or otherwise).

This is a rare occasion when you do not have your calculator or some electronic instrument with such capability. So you wonder whether you can come up with a shortcut to calculate the total number of seats. What would you do without either modern technology or paper and pencil? Chances are good that you are using some form(s) of the distributive property.

If you got the answer before your friend next to you observes that you are about to fall asleep, try to figure it out another way. You try to calculate a quick answer in your head that is smaller than the exact answer, and then you try to figure out a way to correct it. Then do the same for an answer that is slightly larger.

If you haven't already, try to find a precise answer. If you have found a quick mental procedure without explicitly using the distributive property, you probably experienced something akin to what the central character has experienced in the Molière play you have been watching.

So Jourdain has, without realizing it, been speaking prose all his life. If you multiplied the number of rows by the number of seats in each row (25 × 37) in your head, without calculator, pen, or pencil, then you probably used some variation of the distributive property—even if you haven't thought about it in a formal sense defined above.

Now back to calculating the number of seats in the theatre. If you were trying to multiply 25 (the number of rows) by 37 (the number of seats in a row), you might have thought something like this: 25 × 37 is harder to figure out mentally than 25 × 30. That would be 750. Having calculated 25 × 30, what more do you need to figure out 25 × 37? We have to consider the amount left over, 25 × 7.

Alternatively, suppose you noticed not that 25 × 30 would be easy to figure out in your head, but rather 25 × 40. How much would that be? But that answer is too large. What would you have to subtract from that first calculation to get a precise answer?

Now assume that you and a friend decide to go to a fancy restaurant after the play. After the meal, you receive the bill for $74. You decide that a 20 percent tip would be appropriate, but your friend thinks the service was mediocre and that 15 percent would be enough. Calculate in your head how much you would leave in each case. Which was easier? How did you do it? Compare your approaches with someone else. Did either of you seem to intuitively or implicitly use the distributive property?

Chapter 2 will explore some of these essentially algebraic renditions of the distributive property, first looking at a concrete model of the property and then one with graphic appeal. We will also look at some enticing and sometimes perplexing number tricks related to the property.

Some Simple Visual–Kinetic Models
Marbles

Calculating the number of seats in the Molière play visually portrays the distributive property. To sharpen what we have done, choose a small number depicted by marbles in figure 1.1.

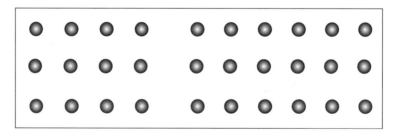

Fig. 1.1. Counting marbles by using two different shortcuts

We can easily figure out the total number of marbles by counting. We can count them individually or in pairs. Alternatively, we can think of them as belonging to one large array, and then count ten across and three down to realize that 3×10 would be the answer. If we envision the array as two collections (3×4 and 3×6), we can calculate the number in each array separately and then add them together. So we can add $(3 \times 4) + (3 \times 6)$ by using a slight variation of the distributive property. We might see that $(3 \times 4) + (3 \times 6)$ is equal to something slightly easier to calculate, especially for those not particularly adept at calculating. That is,

$$(3 \times 4) + (3 \times 6) = 3 \times (4 + 6), \text{ which equals } 3 \times 10.$$

Of course, we have slightly pulled the rug out from under, because we switched the right and left side of what we wrote as the distributive property. It is now in the form

$$(a \cdot b) + (a \cdot c) = a \cdot (b + c).$$

If we move the focus from one side of the equation to the other, the meaning of the property does not change: if $A = B$, then $B = A$ and vice versa. Nevertheless, focusing on the movement of the equation in one direction rather than another is sometimes more helpful. This interesting observation pervades much of mathematical thinking. But this kind of fine point will creep up on us in other contexts later. Simply knowing that two different ways of expressing the same thing yield the same result does not tell us which one we should focus on in a particular circumstance. And in fact, many other ways might exist to view what would be a helpful transformation of an abstract statement (here an equation) in a problem.

Before moving to further conceptions and elaborations of the distributive property, consider a fine point that the theatre example glossed over. We have done more than notice that we can break 25×37 up into two multiplication examples and then add them to get the answer. We could have expressed 25×37 in many ways other than $25 \times (30 + 7)$. As you read further in this book, you will come up with other clever ways of doing problems like this one in your head.

Divided We Stand

Conceptually, the marble model enables us to visualize a concept that we will use in several sections that follow. The marble example shows what it means for one number to divide another—a kind of language we will use in chapters 2 and 3 for prime numbers. So, in figure 1.1, we see that 3 divides 12 and that 3 divides 18. What can we picture when we say that 3 divides 12? One way is to focus on rows, columns, and total numbers of marbles. It means we can take the total number of marbles (12) and break them up so that we have three rows of marbles with the same number of marbles (4) in each. We know then that 3 divides 12 into groups of 4, so that we can find an array of 3 rows and 4 columns (3×4), which equals 12 marbles. Similarly,

3 divides 18 into groups of 6, so that we can find an array of 3 rows and 6 columns, which equals 18 marbles. Combining these two arrays, we can see that if 3 divides one number (18) and 3 divides another (12), then 3 also divides their sum (18 + 12).

What do we mean when we claim that one number divides another? What do we mean when we say that 3 divides 21, or that 3 does not divide 26? Phrasing the question that way does not ask what 21 divided by 3 is, or what 26 ÷ 3 is, but rather what the statement means when we claim that one number (think of the counting numbers for now) divides another. We will return to this matter in chapter 2 from an algebraic point of view when we examine some number tricks.

A Dynamic Approach

I have already suggested one concrete model that illustrates the distributive property for natural numbers—the marbles. Others exist, and one particularly appropriate as a concrete model is the simple balance scale in figure 1.2. It can serve as a mind experiment for many readers, an invitation to construct variations of the distributive property by modifying components.

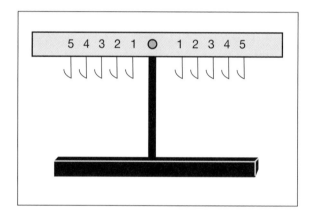

Fig. 1.2. A simple balance scale without weights

Many ways exist to construct such a scale—enabling it to swivel at the middle point on top and calculated so that the length of the scale is the same on the right and left of the swivel. Place hooks so that they are equally spaced from each other and from the center. To begin, take any number of rings and place some of them on one hook and the same number on any other—all on the same side of the balance scale, as in figure 1.3. See what happens as you change the placement of the rings, using the same number of rings on both hooks.

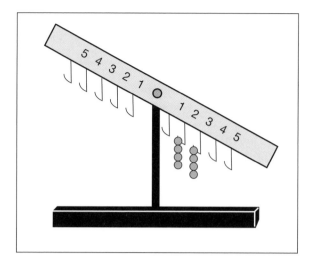

Fig. 1.3. Unbalancing a scale with rings on one side

The scale would now be unbalanced. Now you might play around with where and how many rings you might place on the other side of the scale to balance it (fig. 1.4). How many different ways can you do it?

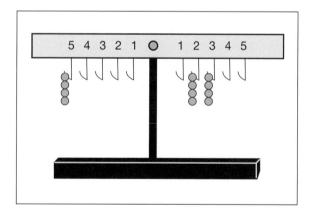

Fig. 1.4. Rebalancing a scale with different numbers of rings on both sides

Some Modifications

We could achieve balance in several ways by placing some rings on various left-hand hooks. It would balance if we put two rings on a hook located ten units to the left;

it would also balance if we put one ring on a hook located twenty rings to the left (though the scale will probably not be that long). It would also balance if we placed five rings on the #4 hook or four rings on the #5 hook.

Using this material, we can create interesting games around such requirements as the desire to balance the scale with as few or as many rings as possible. Anyone playing with a balance beam for the first time will profit by freely experimenting with it. We can then define further requirements so that in solving the balance of both sides of the balance beam, one could depict the distributive property. We could summarize the activity in figure 1.4 by replacing the question mark with a particular number:

$$(4 \cdot 2) + (4 \cdot 3) \qquad = \qquad (4 \cdot ?)$$
(Left side of balance)   (Right side of balance)

We are adopting many conventions here, and seeing why addition and multiplication are reasonable operations to describe hook locations and ring placements might take a while. Several examples suggest perhaps why 5 is a solution for "?" in the above equation. Finally, we can summarize the activity in general by the following:

$$(n \cdot k) + (n \cdot m) = n \cdot (k + m),$$

where k and m refer to the distance of the hooks from the center and n is the number of rings on each hook.

Reflections on the Above Model

The balance beam is essentially a seesaw, or teeter-totter. We could use people themselves on a marked seesaw to generate some of the ideas discussed above. For distributivity, the only slightly awkward requirement is that three people demonstrating the principle be roughly the same weight. Especially if people serve as weights, however, some doubling up could occur, and people would probably be inclined to push the prescribed boundaries to raise issues other than the distributive property. (One might even use more than three people to explain generalizations of the distributive property.)

Dienes (1961, p. 80ff) controversially claims that one needs not only practice with a model or scheme to acquire a new mathematical concept but also to experience multiple embodiments of a mathematical concept to grasp its essentials and avoid fixating on irrelevant features. I suspect that many of us have correctly acquired a concept from one instance, provided that one specific instance can be seen in both concrete and abstract terms. Nevertheless, the concept of multiple embodiment raises some interesting pedagogical questions and is discussed as part of a more general theory of mathematical learning by Dienes.

Some Geometric Thinking

Joining the two arrays of marbles approximates the distributive property in geometric terms. If you break a rectangular figure into two parts by drawing a perpendicular line from one side to the other, we have figure 1.5:

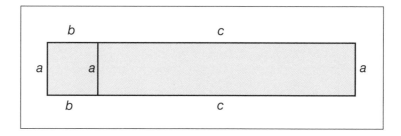

Fig. 1.5. Geometric and algebraic schemes of the distributive property combined

As we indicated with the marbles, we can calculate the area in two different ways, as $a \cdot (b + c)$ or as $(a \cdot b) + (a \cdot c)$. This, of course, is a standard area formula associated with Euclidean geometry. Although you probably dealt with geometric renditions of algebraic formulas in high school, Euclid would not have made the relationship between figure 1.5 and the algebraic rendition. Though Euclid, born about 325 B.C., based many of his proofs on such diagrams, he never used algebraic formulas like that above in his famous *Elements of Geometry*. Why not? This Greek mathematician, born three centuries before Christ, used a straight edge and a pair of compasses but did not have a ruler depicting the numbers that we now comfortably associate with the real numbers. Consequently, he could not actually calculate areas of geometric figures in the way shown here. He never used such simple formulas as "The area of a square with side x is: x^2" or "The area of a triangle is ½ base × height."

Despite such a limitation in formulating ideas of geometrical measures, however, Euclid's ability to state and prove theorems accurately, albeit nonnumerically, is an act of genius. What precisely was he doing that enabled him to think of the rectangles above as depicting a form of the distributive property? How did he ever formulate the famous Pythagorean theorem (dealing ostensibly with areas of squares on the three sides of a right triangle), much less prove it, without using numbers for their areas? We will also explore how the algebraic versus geometric conception of the distributive property might have affected how the property was generalized.

These are historically fascinating matters we will discuss in chapter 7, but until then, we will continue to connect algebraic formulas with geometric figures as we extend and apply aspects of the distributive property.

Illustrating the Distributive Property

Geometric and algebraic ideas are sometimes joined in interesting ways through graphical representations. Look at the graph of a simple linear equation in figure 1.6.

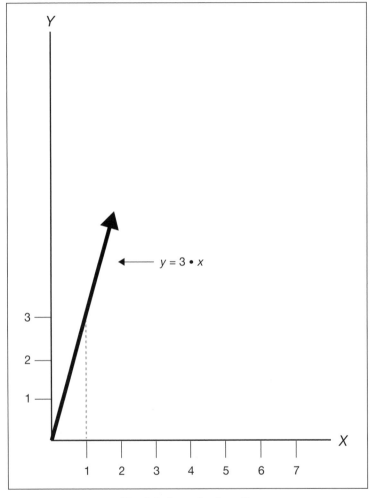

Fig. 1.6. A graph of $y = 3x$

Now consider x values of 2 and 5, and see where they hit the straight line (fig. 1.7).

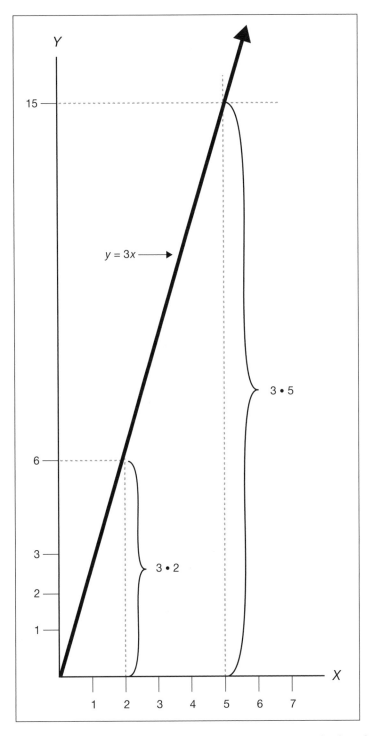

Fig. 1.7. Vertical lines at $x = 2$ and $x = 5$, intersecting the graph of $y = 3x$

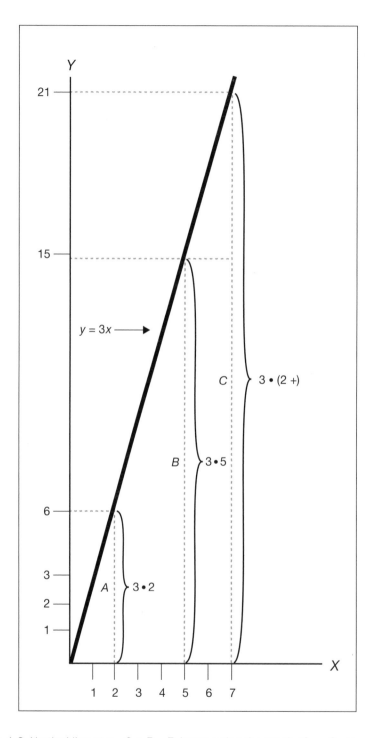

Fig. 1.8. Vertical line at $x = 2 + 5 = 7$, intersecting the graph of $y = 3x$ at $y = 21$

The corresponding heights (y-values) are 6 and 15. Now suppose we locate the point on the x-axis that is the sum of the two lengths we selected on that axis. So we have the new x-value of $2 + 5 = 7$. Where does the corresponding y-value fall (fig. 1.8)?

As the graph shows, it falls at $y = 21$. That is no surprise because we set up the situation so that every value of x would project onto a y-value on the graph so that $y = 3x$.

Interestingly, the y-value for $x = 2 + 5$ (line C on the graph) is the same as the sum of the y-values for $x = 2$ and $x = 5$, respectively (lines A and B). So we could have predicted the length of C by adding the segments A and B. Why is this happening? The length of the y-value at $x = 2 + 5$ is equal to the length of the y-value at $x = 2$ plus the length of the y-value at $x = 5$.

But how were these three lengths calculated? It is not just that $3 \cdot 2$ is one length, $3 \cdot 5$ is the second, and $3 \cdot 2 + 5 = 21$ is the third, but rather that by now the infamous relationship holds as a consequence of the distributive property. So if $y = 3x$ defines the relationship between x and y, then for any two points a and b,

$$3 \cdot (a + b) = 3 \cdot a + 3 \cdot b.$$

If this seems rather humdrum, investigate what happens with other linear equations (more generally, $y = n \cdot x + m$) and see whether this condition still occurs. Then investigate what happens with other relationships. Try the relationship among three such points projected from the x-axis onto the curve $y = x^2$ (fig. 1.9).

If the same relationship that we suggested for straight lines held for the parabola, then not only would C fall on the parabola, but the two y-values (A and B) would also equal the length of the third. That is, $2^2 + 5^2 = 4 + 25$ would have to equal $(2 + 5)^2 = 49$. Here, if you try to relate the heights of the three points projected from the x-axis by adding the x-values of the first two (a and b) to get the third, we would have the lengths of the projections of the first two, a^2 and b^2, whereas that of the third would be $(a + b)^2$.

I have tried to maintain a light tone here, not overemphasizing precision in order to invite you to fill in or ask questions about the role of the distributive property in some graphs and not others. You might want to work through examples or general questions to see what might be true about linearity and the distributive property. For example, could equations besides those of lines exhibit the additive quality for three points projected from the x-axis, as described here? Might such additivity not hold for some geometries? Could lines other than those of linear equations exhibit this additivity status even if the distributive property does not hold?

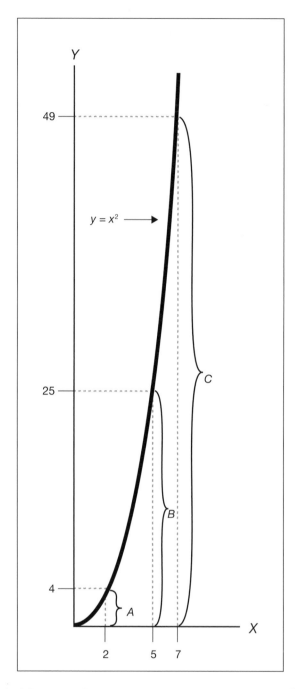

Fig. 1.9. Vertical line at $x = 2 + 5 = 7$, intersecting the graph of $y = x^2$ at $y = 49$

The Distributive Property and the Computer's Last Stand

The distributive property has an interesting feature that you can best understand by imagining that the three numbers *a*, *b*, and *c* are entered into a computer for two different programs. If you had 5, 8, and 17 placed into one program, you might direct the following calculation:

$$5 \cdot (8 + 17)$$

What would the computer do? It would add 8 and 17 and then multiply the sum by 5. Even if the computer did the calculation differently, multiplication would be the last operation performed: $5 \cdot 25$. Acknowledging the distributive property, however, we would program the calculator differently. For $5 \cdot (8 + 17)$, we might tell the computer to calculate the following:

$$(5 \cdot 8) + (5 \cdot 17)$$

What operation would the computer perform last now? It would first calculate $5 \cdot 8$ and then $5 \cdot 17$; finally, it would add the two products. The last operation now would be addition rather than multiplication.

So one intriguing way to conceive of the distributive property is to envision it as a statement that two different computer programs for three random numbers yield the same results. In one program, however, the last step is addition, whereas in the other, it is multiplication.

A Source of Confusion

The distributive property is harder than the others (in the section below) for students to understand at early stages. It may be the most difficult to apply to new situations, is easily mixed up, and is most readily forgotten. Students might spend considerable time on drill and practice with the property, only to conclude that the following is an instance thereof:

$$2 \cdot (5 + 7) = 2 + 5 + 7$$

Students much more readily grasp the other number properties depicted below (even without labeling them). These, plus the distributive property, are components of systems we will use. Chapter 6 has a more careful statement and analysis of these properties.

15

The Beginnings of a "Field" Day

Here are other properties we will be using:

- Closure under addition: If a and b belong to a specified system, then $a + b$ belongs to the same system.

- Closure under multiplication: If a and b belong to a specified system, then $a \cdot b$ belongs to the same system.

- Commutative property for addition: $a + b = b + a$.

- Commutative property for multiplication: $a \cdot b = b \cdot a$.

- Additive identity: $a + 0 = a$.

- Multiplicative identity: $a \cdot 1 = a$.

- Associative property for addition: $a + (b + c) = (a + b) + c$.

- Associative property for multiplication: $a \cdot (b \cdot c) = (a \cdot b) \cdot c$.

What makes the distributive property more confusing? For one, of all the elementary properties (axioms), only the distributive property involves two operations in one statement. Other reasons may exist; after you finish this book, reexploring the issue would be worthwhile.

Because the distributive property involves two operations, we can explore some fascinating questions—not only of a mathematical nature but also of a metamathematical sort. These questions probe the nature of mathematics and of mathematical systems and their properties. In chapter 6, we will analyze them, focusing on the distributive property. We will explore questions such as this: in a system that captures the real numbers, can I know beforehand whether applying the distributive property to a particular theorem will be necessary?

We turn now in chapter 2 to several playful extensions and connections with the distributive property.

References

Avital, Schmuel. "The Plight and Might of the Number Seven." *Arithmetic Teacher* 25 (February 1978): 22–24.

Dienes, Zoltan P. *Building Up Mathematics*. London: Hutchinson Ltd., 1961.

Eastis, David M. *7: The Magical, Amazing, and Popular Number Seven*. Avon, Mass.: Adams Media, 2011.

Hoffer, Alan. "What You Always Wanted to Know about Six but Have Been Afraid to Ask." *Arithmetic Teacher* 20 (March 1973): 173–80.

Hohfeld, Joe, and Lynn Schwandt. "Six Is a Fascinating Number." *Arithmetic Teacher* 22 (April 1975): 269–70.

Jones, Philip Dwight. *The Middle-Class Gentleman* [English trans. of Molière's *Le Bourgeois Gentilhomme*]. 2008. http://www.gutenberg.org/files/2992/2992-h/2992-h.htm.

Newman, James. "Srinivasa Ramanujan." In *The World of Mathematics*, edited by James Newman, pp. 368–76. New York: Simon and Schuster, 1956.

Playfulness

Rain Man

In the 1988 movie *Rain Man*, when the waitress accidentally dropped a box of toothpicks, protagonist Raymond Babbitt looked down and—apparently without thinking—muttered, "82, 82, 82." She looked at the box of toothpicks and noticed that it contained 250 of them. She looked in the box and was astounded that four toothpicks had not fallen out. Raymond (an adult in the movie, brilliantly played by Dustin Hoffman, who won an Academy Award for the role) had been diagnosed with autism, a condition often associated with lacking understanding of social skills and a delay in acquiring language. Autism occurs along a spectrum, and effective treatments to minimize the condition are now available. Some autistic people have extraordinary competence in mathematics and science.

Those closest to Raymond perceived him as mentally retarded and labeled him an idiot savant. In addition to the feat above, he could calculate 4343×1234 in his head without formal coaching. He showed such competence despite his inability to cope with apparently simple acts of communication. When asked, "If you had a dollar, and you spent fifty cents, how much money would you have left?" he answered, "About seventy cents."

Here's another nonmathematical interchange between Raymond and his brother, Charlie, who queried him about pancakes Charlie planned to make for him: "Ray, we've got blueberry, buckwheat, all flavors. What kind do you want?" Ray's answer: "Pancakes." In response to Ray's answer, Charlie asks again, "I know, but what kind?" Ray answers once more, "Pancakes."

How Raymond could function at such a high level mathematically when he could barely communicate coherently in most social contexts is a mystery. This chapter explores procedures and tricks people use to pull off some calculations quickly and in their heads instead of using electronic devices or writing equipment. Ray or other savants of that ilk probably never explicitly used these clever shortcuts, but these tricks can entice and surprise people who observe such calculations. This exploration also raises our awareness of the delicate interplay between competencies that are built in, or expressed intuitively, and those acquired through explicit teaching strategies.

"Prime" Concern

Studying prime numbers from a historical point of view is fascinating. In a sense, prime numbers are building blocks for the counting numbers, for nonprime numbers can be decomposed into a product of numbers that are all prime. By definition, a number is prime if it has exactly two different numbers that divide it (without remainders). So the first few primes are 2, 3, 5, 7, and 11. Why are 9 and 4 and 1 not prime?

We barely define the concept before we come up with a cascade of questions. What questions can you come up with? A few that are relatively innocent sounding could actually be ones that stumped the mathematics community for years. Here are some questions that might be on your list:

- How many primes exist? Can a formula generate these primes? All of them? An infinite number but not necessarily all?

- Do the primes distribute themselves in some interesting manner?

- Can we determine, without laborious calculation, whether a number is prime?

- How do prime numbers connect with properties of the counting numbers?

Euclid resolved one of the earliest questions—how many primes exist—more than 2000 years ago. The question is reasonable, because if you start counting the first hundred primes or so, and then explore how many occur in the next hundred and the following hundred, they tend to become much sparser. Chapter 7 will examine Euclid's answer to this question. There we will also look at several other historically interesting questions, ones that deal with the representation of even numbers in terms of primes (Goodbach's conjecture of 1742) and with the occurrence of twin primes (those that are two numbers apart).

The rest of this chapter, however, will use primes together with some surprising number tricks to expand formulations and extensions of the distributive property. To begin that exploration, this chapter continues with a modest question that may appear unexpected.

Is 3599 a Prime Number?

The difficulty in determining whether 3599 is prime is not only the number's size. For example, we know immediately that 3598, 3600, 3602, 3604, and 3636—being divisible by 2—are all nonprime, even though they are similarly large. Furthermore, 3605 is also composite, because it must be divisible by 5. The only challenging numbers, then, are odd numbers not ending in 5: 3597, 3599, 3601, 3603, 3607, and 3609.

Well, is 3599 prime? To find out, we could systematically divide it by possible candidates. We need not try dividing it by any even number or by a number ending in 5. But this process can still be laborious. Can you guess how many you would have to try? It would be a lot. Intuition might tell us that testing by numbers greater than 3599 would be unreasonable. But why? Why are we so sure that no numbers greater than the number being tested might divide it? At first, we might conclude that only a madman would consider such a possibility. Presuming that madmen and geniuses share important qualities, think about why this is so obvious.

We could also search for a more efficient answer by noticing that 3599 is almost 3600—a perfect square: $3600 = 60^2$. So what? Well, it misses by 1, so we really want to know whether $60^2 - 1$ is divisible by something other than 1 and itself. If we can express the number as a product of numbers (other than 1 and itself), the number would not be prime. You may see something here that is a bit of a leap, but if not, think about what 60^2 has that 1 does not. The number 1 is not a square—or is it? Try to represent 1 as a square. It is the only nonzero number that is its own square.

So how can we look at $60^2 - 1^2$? If you have not thought about such matters algebraically, perhaps you can think of the difference of two squares in nonalgebraic terms. That we refer to each value as "square" suggests that we could use some geometrical formulation of the problem, as we did in the rectangular model from chapter 1. Though many variations exist, the sketches in figures 2.1 and 2.2 occurred to me (with a representing 60 and b representing 1).

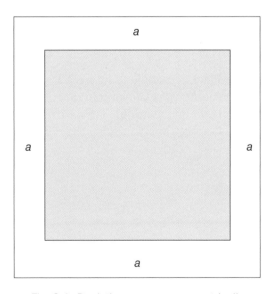

Fig. 2.1. Depicting a square geometrically

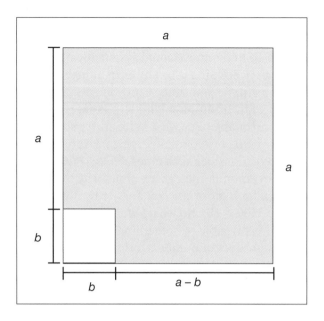

Fig. 2.2. Deleting a small square from within figure 2.1

If we found two numbers (other than 1 and 3599) that multiply to give 3599, we could show that 3599 is not prime. Geometrically, we could achieve this by coming up with a rectangle that has two such numbers as the sides. So how do we break up figure 2.2 so that it consists of a rectangle with the same area as the shaded area of figure 2.1? Look at figure 2.3. We have separated the desired region into two rectangles. Although we could add up their two regions algebraically, let's first see what happens if we strive to find only one large rectangle composed of the two smaller ones in figure 2.3.

Almost voilà. Take the black rectangle in figure 2.3, rotate it 90 degrees, and place it on top of the gray rectangle, as shown in figure 2.4. We thus have two rectangles that combine to form one rectangle with an area equal to that of the shaded region of figure 2.2 (appropriately missing the region of the little square at the bottom).

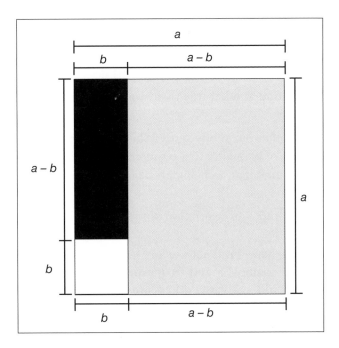

Fig. 2.3. Seeing figure 2.2 as two rectangles

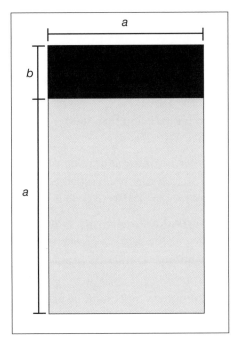

Fig. 2.4. Transforming two rectangles into one having the same area

Several Bonuses

The above geometric scheme has several benefits that have surfaced or soon will:

- We have determined whether 3599 is prime.
- We have found another way to answer that question without laborious calculation.
- We have somewhat subtly shifted the original, purely arithmetic conception of primality to a geometric one. That is, we can define a counting number, n, as being prime not only in arithmetic but also in geometric terms.
- We have found a new variation of the distributive property.

What have we done here? I did not use an already known procedure to determine whether 3599 is prime. I found an approach that works in this case, and I wondered how far I could apply it to other cases. This approach to determine primality has some limitations. But along the way, I found an expansion of the distributive property that I knew beforehand, but not one from such an interesting geometrical perspective, and one that we can generalize beyond its use in determining primality.

This analysis shows the richness of intertwining the specific and the general, of finding new ways to relate algebraic and geometric thinking. What observations propelled me to investigate whether 3599 is prime?

1. A metaphor we will see in the next chapter: 3599 has an interesting personality. I saw 3599 as "striving" to be something else: the square number 3600, which equals 60^2. So $3599 = 60^2 - 1$.

2. I made a leap from observing that 3600 is a perfect square: 1 has the same property ($1 = 1^2$).

3. I wondered whether invoking square numbers might suggest a geometric rather than a strictly algebraic scheme. I investigated figures 2.3 and 2.4, finding a way to depict 3599 as a 59×61 rectangle.

4. Now we have a second rectangle with dimensions other than 3599×1, with the same area as the first—a powerful visual scheme that lets us determine that a number cannot be prime.

Think about other large numbers that could be expressed like this. Which of the following might work? 3597? 3591? 6989? 6999? 7991? 7893?

If we move from an examination of the connection between algebra and geometry with a focus on primes, let's see how this geometric manipulation sheds interesting light on algebraic thinking in more general terms. Look again at figure

2.3. Here we have shown how a large square with a small one removed from within can be cut and pasted in a way that yields a differently shaped region with the same area, depicted in figure 2.5. This is summarized below.

$$a^2 - b^2 = (a - b)(a + b) \tag{1}$$

Furthermore, we can derive this property algebraically from the original distributive property by thinking of $a - b$ as being one number, x. For then

$$(a-b)\cdot(a+b) = \underbrace{x\cdot(a+b)}_{i} = \underbrace{x\cdot a + x\cdot b}_{ii} = \underbrace{(a-b)\cdot a + (a-b)\cdot b}_{iii} \tag{2}$$
$$= \underbrace{(a^2 - b\cdot a)}_{iv} + \underbrace{(a\cdot b - b^2)}_{v} \text{ (from iv + v)}$$
$$= a^2 - b^2$$

As the songwriter/performer/satirist Tom Lehrer quipped in his "New Math" song of the 1970s, "You ask a silly question, and you get a silly answer!" Wait, what was the question again? If you are used to algebraic arguments, this will look like no big deal. If not, it may appear to be gibberish. Either way, though, it may look tortuous. It could be worse because we have left many steps out of the argument. We have used the properties mentioned in passing in the chapter 1 section "The Beginnings of a 'Field' Day"—something we will use more fully in chapter 6 (see fig. 6.3).

However, we have used the distributive property with some degree of finesse in the above algebraic justification. Though many of you familiar with the algebraic manipulations will quickly see what is going on, others may appreciate some explanation. The use of x in part (i) of equation (2) allows us to consider $(a - b)$ to be *some* number without paying attention to what that number is. In moving to (ii), we then focus on expanding $x \bullet (a + b)$ by using the distributive property. In moving from (ii) to (iii), we focus explicitly on x as $(a - b)$. In moving to (iv) and (v), we have used a slight variation of the distributive property:

$$[u \bullet (v + w) = (v + w) \bullet u]$$

Numbers Ending in 5

An interesting mental trick involves figuring out the square of numbers ending in 5. Before exploring it further, calculate, any way you wish, the square of two-digit numbers that end in 5. For the first six we get the following:

$$15^2 = 225$$
$$25^2 = 625$$
$$35^2 = 1225$$
$$45^2 = 2025$$
$$55^2 = 3025$$
$$65^2 = 4225$$

Part of the pattern is obvious—what we get for the last two digits. What can we say about the first digit of each number that we square? Well, look at what we have so far: 1 yields 2; 2 yields 6; 3 yields 12; 4 yields 20; 5 yields 30. What might 7 (as in 75) yield?

To get started, it would help to appreciate something that may be so obvious that it is taken for granted. We are dealing with numbers expressed in base ten. For example, the number 65 is shorthand for a base-ten number. For a clue: how do we pronounce "65"? We call it "sixty-five": the number is a shorthand for $6 \cdot 10 + 5$. If t depicts any of the first digits of the array of numbers, then we can depict the numbers as

$$10 \cdot t + 5, \qquad (3)$$

where t stands for each digit in the pattern above. To say that we are seeking the square of each number suggests that we might be looking for another way of depicting:

$$(10 \cdot t + 5)^2 \qquad (4)$$

Many of you may know how to express the above square differently (and many ways exist). But if we focus on the concept of square, we might be tempted to think of the number as being described geometrically as we did in figures 2.1–2.4. That is, how could we think of $(a + b)^2$ in relation to geometric squares? Many sketches might be useful. We might begin with figure 2.5:

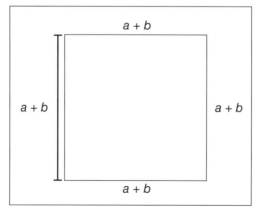

Fig. 2.5. Geometric representation of the sum of two numbers squared

Figure 2.6 is a refinement of figure 2.5, showing us two geometric squares in relation to each other.

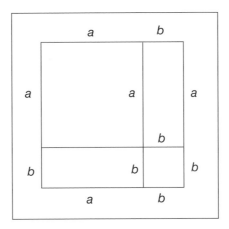

Fig. 2.6. Breaking figure 2.5 into four components

This may be a bit of a leap if you have not transformed algebraic terms before, but now we have a large square depicting $a + b$ as a side. Now, what is $(a + b)^2$ composed of? Reexamine figures 2.1–2.4. Here we have the large square composed of four regions. We have two squares, a^2 and b^2, and then two rectangular regions, each with area $a \cdot b$. So, we have a nice geometric rendition of the following algebraic formulation:

$$(a + b)^2 = a^2 + 2a \cdot b + b^2 \tag{5}$$

Now we have two interesting extensions of the simple distributive property, one for the difference of two squares [equation (1)] and one for a square whose length is the sum of two lines $(a + b)$ [equation (5)]. We can prove this algebraically, as we showed in equation (2) for the difference of two squares, but I leave that for you to do or to discuss with someone who may have worked through such derivations before.

How does this new rendition of the distributive property show how you can calculate in your head the squares of two-digit numbers ending in 5? Looking at our depiction of the square of any two numbers, as in (5), and at our depiction specifically of the square of two-digit numbers ending in 5 above—and expressed in base ten—we have:

$$(10 \cdot t + 5)^2 = (10 \cdot t)^2 + [2 \cdot (10 \cdot t \cdot 5)] + 5^2 \tag{6}$$

If we simplify (in a way to make obvious what may be the very cute shortcut)—after all this sweat—we get:

$$100 \cdot t^2 + 100 \cdot t + 25 \tag{7}$$

If we go one step further, we finally have an algebraic rendition for the pattern of squares ending in 5:

$$100 \cdot [t(t + 1)] + 25 \tag{8}$$

That little inclusion of $t(t + 1)$ involves another subtle use of the distributive property. We can reexpress $100 \cdot t^2 + 100 \cdot t$ in (7) as $(100t) \cdot t + (100t)$, which by the distributive property we can express as $100t \cdot (t + 1)$.

So where are we? If you look at any two-digit number expressed as $10t + 5$ and express it as a square, you end up with (8) as a shortcut. How is it a shortcut? The $t(t + 1)$ brings out in the open not only the solution but also the statement of how to perform the shortcut. That is, just multiply the first digit of the two-digit number ending in 5 by one more than that digit, and you get the square of the number by appending 25 to the result. So to get 75^2, find the digit after 7 (which is 8), and prefacing 25 by $7 \cdot 8$ gives you the answer of 5625! (not factorial).

Though many of you may have derived (1) and (5) before (if not an excruciating headache), these extensions may not have been done as an act of interrelating algebraic and geometric thinking. More important, this approach portrays the geometric development through the use of one geometric scheme of relating squares to each other in each extension. You can find a briefer sketch of some of the idea related to these past two subsections in Brown (1987).

Our next number trick depends heavily on the distributive property and makes use of our base-ten number system. That number is associated with the proverbial cat.

Nine Lives (Double Entendre: A Verb and a Noun)

Using their experience with odd and even numbers, most people would predict the nature of the sum of two even numbers, the sum of two odd numbers, and the sum of an array of even and odd numbers. The various models of chapter 1 would make it easy to come up with conjectures, if not proofs.

Proving the nature of combinations made by evens and odds takes little more than depicting even numbers (expressible as $2 \cdot n$ for any integer n) and odd numbers (expressible as $2 \cdot n + 1$) in a convenient format. Alternatively, as chapter 1 discussed, we could depict numbers as even if they are divisible by 2 (meaning that a number m is even if a number n exists so that $2 \cdot n = m$).

Determining whether a number is even or odd (as opposed to combining such numbers) takes no more than examining the last digit. If it is 0, 2, 4, 6, or 8, the entire number is even regardless of what digits precede it. Otherwise, the number is odd.

Familiarity with such numbers can make determining how and why things work easier. Such familiarity, however, may mask the tools that support our beliefs. For one thing, in determining that a number is even or odd, we are inclined to overlook that the numbers are usually expressed in base ten. If numbers were expressed in base two, for example, then "2" would never even appear as a digit, and yet the parity of a number (even vs. odd) would still make sense.

To gain a clearer idea of the distributive property's role and of the power of the system we use to denote numbers, we will explore some questions like the above for the number 9. Using the approach in chapter 1, we can analyze what happens when you combine multiples of 9. You might conjecture that any time you add two numbers divisible by 9, their sum is also divisible by 9. Try a few more and then ask how you would prove it. We can use the marble model as we did in chapter 1.

An Algebraic Approach

What does it mean to say that 9 divides a number "evenly"? As we discussed in chapter 1, it means that the number can be expressed in the form $9 \cdot n$ for some counting number n. So if 9 divides each of two numbers, the numbers are of the form $9 \cdot n$. Now we are asking what you might say about the nature of their sum. An obvious start would involve getting the sum of the two numbers. It would be expressed as $9 \cdot a + 9 \cdot b$ for numbers a and b.

Now what can we say about $9 \cdot a + 9 \cdot b$? We can say a lot of things, but a small leap—repeating such thinking in the previous chapter—involves using the focus of this book. Thus, $9 \cdot a + 9 \cdot b = 9 \cdot (a + b)$. What is the advantage of so reexpressing $9 \cdot a + 9 \cdot b$? Because the sum of any two whole numbers a and b is another whole number (that is, the property of closure under multiplication), $9 \cdot (a + b)$ is of the form $9 \cdot n$, and thus it is another number also divisible by 9.

The above demonstration is a simple proof that depends on "field" properties we referred to in chapter 1:

- The meaning of divisibility by 9
- The distributive property
- The closure property for addition and multiplication for whole numbers

When Nine Stands Alone

Determining whether any number standing alone is even or odd is easy. You can easily tell whether any number is divisible by 5 without fancy calculations. You might want to think through what you can tell about the divisibility of other numbers just by looking at them. Let's look at the situation with the number 9.

Have you wondered before whether a specific number is divisible by 9? (If you are of Germanic descent, then your answer to the question, "Have you played around much with nines?" might very well be "nein.") We will soon present a feat that will probably surprise you. Will looking at the last digit of the number tell you whether it is divisible by 9? Look at the following to see whether that approach works: 9, 19, 49, 99.

Might some other digit appear as a last digit that would determine whether a number is divisible by 9? Try the following: 21, 41, 81, 321, 351.

See whether another potential final digit can determine divisibility by 9. In both cases above, the answer appears to be "sometimes yes; sometimes no." Is that the sort of test we might be seeking? So far, we have not found a last digit that would always yield a number divisible by 9. Can you find one that would never work?

A Number out of the Blue

I have chosen a relatively large number out of the blue, but after leaving it in the sun for a while, the fourth digit has been obliterated:

$$563,?17$$

Question: What is the missing digit of my number? Answer: It's a silly question. The missing digit of course could be anything from 0 to 9. Can we make the question more interesting? Try to raise the question to a more challenging level that requires more than observing that any of ten digits might work. Suppose you knew that I was thinking of an odd number. Or an even number. Would that affect how you might answer the question?

With just a little reflection, the question might still not lead to a great deal of payoff. Now suppose we transform the question into one that would be more of an eye-opener. Here is one candidate: what might be the missing digit if I suddenly recalled that the number I had selected was divisible by 9?

Here is one that you could explore. You might start inserting all the digits from 0 to 9 until you come up with a number divisible by 9. That approach might be laborious. So would looking at all the numbers in the times table for 9. But it might be easier if you did a different sort of digging of your own. Without focusing on missing digits, look at the numbers in the 9 times table to see what might be interesting about the numbers. List the first five or ten numbers that are multiples of 9. Here are a few others that are also multiples of 9 that are a bit larger: 162, 171, 180, 189, 198, 207, 702, 720, 216, 612, 261, 621, 5328, 3528, 8253, 8235.

First check that they are all divisible by 9. Does anything stand out with all the numbers you have collected that are multiples of 9 (or, equivalently, divisible by 9)? Does that feature work if you select some numbers that are definitely not divisible by 9?

Before seeking some sort of verification, plug in the digit that you calculated to be the missing one for the "?" inserted above—assuming you knew that the number is divisible by 9. List others if there was more than one possibility.

563,?17

Once you have found the (or a) digit that works, look at the following array of numbers formed merely by rearranging the digits in this number (always plugging in the same digit you selected for the "?" in 563,?17).

7?3,651

?73,651

361,57?

7?6,135

5?3,617

517,6?3

176,?35

Many more permutations of these digits exist. How many of these numbers were divisible by 9? Just for kicks, choose another digit than the one you arrived at. Insert it into the spots for each number above.

You perhaps have come up with several conjectures, but one that may be a surprise is arrived at by adding all the digits for any of the above numbers. Try it for all the multiples of 9 that you collected. What do you observe if you add the digits of any number divisible by 9? With this hint, could someone else help you figure out whether a simple way exists to determine divisibility by 9?

Among the many good conjectures you may have come up with, we will focus on the following. Look at the algebraic analysis and see whether it works for a conjecture that might be different from the following:

If the sum of the digits of a number is divisible by 9, then the entire number is also divisible by 9.

Return to an Algebraic Formulation

Looking back at the approach we used to explore the nature of squares of numbers ending in 5, we realize that any three-digit number made of digits abc is really a shortcut (in base ten) for a number that can be expanded as $100a + 10b + c$. If we want to now express a three-digit number and focus on the sum of the digits $a + b + c$, how do we incorporate that into our formulation of the problem? We can begin by rewriting the original number $100a + 10b + c$ so that $(a + b + c)$ is part of the entire number. So any three-digit number can now be expressed as follows:

$$[99a + 9b + (a + b + c)]$$

Then if $(a + b + c)$ is divisible by 9, we can express that portion as $9 \cdot k$ for some k.

So, if the sum of the digits $(a + b + c)$ can be expressed as $9 \cdot k$, the number we are investigating is in the following form:

$$99 \cdot a + 9 \cdot b + 9 \cdot k$$

Now what pops out at you? The entire number is now obviously divisible by 9, as a modest extension of the distributive property suggests.

I conclude with one more problem that depends on the sort of machinery we used for the divisibility-by-9 problem. I received it recently from a friend, Sanford Berens, whom I have known since childhood. He enclosed the following message just before starting the problem: "Here is a math trick so unbelievable that it will stump you. Personally I would like to know who came up with this and why that person is not running the country." Give it a try on your own. I offer my solution at the end of the chapter in case you would like a hint in getting started. Good luck on becoming a national leader.

Unbelievable Math Problem

1. Grab a calculator (you won't be able to do this one in your head).
2. Key in the first three digits of your phone number (not the area code).
3. Multiply by 80.
4. Add 1.
5. Multiply by 250.
6. Add the last 4 digits of your phone number.
7. Add the last 4 digits of your phone number again.
8. Subtract 250.
9. Divide by 2.

Do you recognize the answer? Amazing! All right, you math geniuses, here you go.

Many more astounding number feats and variations depend significantly on the distributive principle. By slightly shifting the domain of numbers from the natural numbers to a subset within it—1 together with the set of even numbers—we will end up wondering how we ever took for granted expectations such as "Every fraction can be reduced to lowest terms in only one way." In doing so, we will see how much more powerful the prime numbers are in the set of natural numbers than we ever imagined. We will discuss some of these features in chapter 7 in the section "Primes: Getting Even."

For an almost end to chapter 2, however, let me highlight an unusually talented mathematician, Arthur Benjamin, who performs what appear to be miracles with numerical calculations (Benjamin and Shermer 2006). You might delight in seeing him perform in several YouTube videos; simply put his name in a search engine. Check out a video of him, "Arthur Benjamin does 'Mathemagic,'" at http://www.ted.com/talks/arthur_benjamin_does_mathemagic.html.

Finally, in case you have difficulty getting started with the "Unbelievable Math Problem" and would like some help, here is something that might smooth the way. It is long but conceptually straightforward.

Notice where the distributive property is used explicitly or implicitly.

As a start, denote the seven-digit number *abcdefg* as a base-ten number as follows:

$$(a \cdot 10^2) + (b \cdot 10^1) + c$$

Multiply by 80 and add 1:

$$(a \cdot 8 \cdot 10^3) + (b \cdot 8 \cdot 10^2) + (c \cdot 8 \cdot 10^1) + 1$$

Now multiply by 250 (here's a big step; when you multiply 8 by 250, you get $2 \cdot 10^2$).

So, we now have this:

$$(a \cdot 2 \cdot 10^6) + (b^2 \cdot 2 \cdot 10^5) + (c \cdot 2 \cdot 10^4) + 250$$

The rest should be a bit laborious but not difficult to complete.

References

Benjamin, Arthur, and Michael Shermer. *Secrets of Mental Math: The Mathemagician's Guide to Lightning Calculation and Amazing Math Tricks.* New York: Random House, 2006.

Brown, Stephen I. *Some Prime Comparisons.* Reston, Va.: National Council of Teachers of Mathematics, 1987.

CHAPTER 3

Multiplication Revisited:
A New Algorithm

**A Journey
of Striving**

This chapter emerged from an essentially metaphorical observation based on noticing something "catchy" as I doodled with the numbers in figure 3.1. The endeavor became more significant than doodling.

$$1 \cdot 3 = 3$$
$$2 \cdot 4 = 8$$
$$3 \cdot 5 = 15$$
$$4 \cdot 6 = 24$$
$$5 \cdot 7 = 35$$

Fig. 3.1. Doodling with numbers (adapted from Brown 1975, p. 546)

Without much thought or analysis of where it might be headed, I noticed that each product was more interesting than I had anticipated.

Soon my awareness became more intense as I focused on the numbers on the right-hand side of the equation: 3, 8, 15, 24, 35. I had met this kind of progression before. The differences between the successive right-hand vertical terms in figure 3.1 were not constant. Rather, they followed the progression 5, 7, 9, 11 (for example, $8 - 3 = 5$; $15 - 8 = 7$; $24 - 15 = 9$).

I was tuned into making this type of observation because such differences (arithmetic differences) signal something about the kind of equation (quadratic) that might generate the original pairs. I noticed an even more remarkable quality about the numbers on the right-hand side. I saw them not as what they were but rather as what they strove to become. With just a little squinting, these numbers were all almost something other than what they appeared to be—and furthermore they all missed being that something else by the same quantity: one unit. Each number on the right-hand side is almost a perfect square. As any of you who read chapter 2 may have noticed, I became intrigued with the metaphor of numbers "striving" to become squares (fig. 3.2).

1 • 3 almost equals 4

2 • 4 almost equals 9

3 • 5 almost equals 16

4 • 6 almost equals 25

5 • 7 almost equals 36

Fig. 3.2. Numbers in figure 3.1 striving to become squares
(adapted from Brown 1975, p. 547)

I was tempted to move in two directions simultaneously. One impulse was to modify the multipliers in the original list of figure 3.1 to determine empirically what the consequences of such modifications would be. If something so remarkable could occur with pairs that differed by 2, what might happen if they differed by some other fixed quantity? Would a "striving" metaphor still exist for me to pursue? A second impulse was to try to understand why this was happening. My inclination was to seek some sort of algebraic or geometric understanding—the reason being that multiplying pairs seemed to be speaking to me about areas of rectangle-shaped figures. My desire to seek something geometric was motivated in part by the realization that, although algebra might offer some formal justification, it might not enable me to "see" what might have motivated the algebraic moves. In line with my first impulse, I noticed the results in figure 3.3.

1 • 5 = 5

2 • 6 = 12

3 • 7 = 21

4 • 8 = 32

Fig. 3.3. What happens to striving if the original pairs differ by 4
(adapted from Brown 1975, p. 547)

What happened to the "almost squares" on the right-hand side of the equation now? Many ways exist to see what might be going on. Could I somehow maintain the concept of striving to be squares in figure 3.3? Two possibilities occurred to me (figs. 3.4 and 3.5).

1 • 5 = 5 [4 off] 1	1 • 5 = 5 [4 off] 9
2 • 6 = 12 [8 off] 4	2 • 6 = 12 [4 off] 16
3 • 7 = 21 [12 off] 9	3 • 7 = 21 [4 off] 25
4 • 8 = 32 [16 off] 16	4 • 8 = 32 [4 off] 36

Fig. 3.4. Striving to maintain squares

Fig. 3.5. Different striving that maintains squares

In each case, I forced the issue of creating squares on the right-hand side. In figure 3.4, I created a sequence of squares beginning with 1 as the first square, 4 as the next, 9 as the next, and so forth. We have a nice, predictable sequence of square numbers in figure 3.4, but they do not increase in the same fashion as depicted in figure 3.2. They depart systematically from the actual product by 1, 4, 8, 12, and 16. That result is worth exploring, but it was not as enticing as the results of figure 3.2.

Figure 3.5 is another story. Here in every case the actual product and the square it strives toward differ by the same quantity. Each square in figure 3.5 is four more than the actual product.

Sticking With It a Bit Longer

Figures 3.2 and 3.5 are interesting for unveiling a degree of regularity, but have they any further significance? You might wish to continue seeking some pattern of striving to be squares that resembles what we found in figures 3.2 and 3.5. Of course, we could start off with any two numbers to multiply, but having begun with 1 • 3 = 3 in figure 3.1 and 1 • 5 = 5 in figure 3.5, I was inclined to explore what happens with a sequence of products beginning with 1 • 7 = 7. We would then have figure 3.6.

1 • 7 =   7
2 • 8 = 16
3 • 9 = 27
4 • 10 = 40

Fig. 3.6. Pairs that differ by 6

Holding in abeyance for a while longer, can you come up with squares that each product is striving toward? As before, many ways exist to answer this question, but if you assume that we are looking for squares that differ from the product in each equation by the same quantity (as in fig. 3.3 for the 1 • 3 sequence and in fig. 3.5 for the 1 • 5 sequence), what would you select? That is, 1 • 3 strove for 4 as the square number; 1 • 5 strove for 9. If you took a good intuitive guess for the squareness candidate for 1 • 7 = 7, what would it be? If 4 is the first square in figure 3.2, and 9 is the first square in figure 3.5, then it might not be a bad hunch that the next possible square to investigate would be 16. That number would be 9 more than the first product in figure 3.6. We would then have the following observation for striving to be squares in figure 3.7.

$$1 \cdot 7 \; = \; 7 \;[\text{add } 9] \;\; 16$$
$$2 \cdot 8 \; = 16 \;[\text{add } 9] \;\; 25$$
$$3 \cdot 9 \; = 27 \;[\text{add } 9] \;\; 36$$
$$4 \cdot 10 = 40 \;[\text{add } 9] \;\; 49$$

Fig. 3.7. A good hunch for what squares figure 3.6 might be striving toward
(adapted from Brown 1975, p. 547)

Payoff of Patience

You may want to glance through what we have done so far. If it is puzzling, or if you feel that you have been pulled by a string through the nose in the dark, there is good reason for it. I have not yet clearly formulated a problem but rather have tried to engage you in following along with my musings. If you have these feelings, you have patiently followed the somewhat cryptic path I was on before I realized that it would head somewhere worthwhile.

I finally came close to formulating a problem when I stopped and drew a sketch of each product. Instead of looking at the two numbers to be multiplied, I focused on their midpoint in relation to the two numbers, as in figure 3.8. The motivation was that I noticed that the striving numbers were 4, 9, and 16.

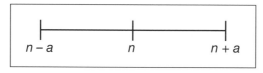

Fig. 3.8. Focusing on the midpoint of two numbers to be multiplied

When I started with 1 • 3, the midpoint was 2. When I started with 1 • 5, the midpoint was 3. When I started with 1 • 7, the midpoint was 4.

So, if the midpoint was 2 for the sequence beginning with 1 • 3, we could think of 1 • 3 as:

$$(2 - 1) \cdot (2 + 1) \tag{1}$$

If the midpoint was 3 for the sequence beginning with 1 • 5, we could think of the product as:

$$(3 - 2) \cdot (3 + 2) \tag{2}$$

If the midpoint was 4 for the sequence beginning with 1 • 5, we could think of the product as:

$$(4 - 3) \cdot (4 + 3) \tag{3}$$

You can see how this follows through with the other entries in figures 3.2, 3.5, and 3.7.

Now where is all this obsession with squares coming from? Look at (1)–(3). It did not begin until I happened to observe that all the numbers of figure 3.1 were almost squares (missing by 1). I wondered whether, if I modified the sequence as in figure 3.3, I could end up with "almost squares" again (so that all products in fig. 3.5 could become squares by adding the same "magic number" to the product). I found that worked by adding 4 to each product in figure 3.5. What was the magic number for figure 3.7? By adding 9 to each product, I created an array of numbers, each a square.

Formulating the Problem

After a somewhat tortuous trek that involved following my musings, we are now ready to ask the question and are almost ready to pose it clearly as a problem.

Start by getting the midpoint of two factors to be multiplied. If n is the midpoint, then the two factors could be expressed as $n - a$ and $n + a$. If you multiply them, what do you get? That is, how can you express the product of $(n - a) \cdot (n + a)$? Wait a minute: we actually created a geometric model for this problem in chapter 2 when we tried to figure out whether 3599 is a prime number. If you have not read that chapter yet, look at the discussion surrounding figures 2.3 and 2.4. We showed there that 3599 is not prime because we can express it as $(60 - 1) \times (60 + 1)$. We knew that because the product is a result of the difference of two squares: $60^2 - 1^2$.

It looks like we have created an alternative algorithm for computing some multiplication problems, one quite different from the standard procedure. Furthermore, the algorithm reduces multiplication to something simple: finding midpoints between the two factors to be multiplied and using squares that are intimately involved in the calculation.

Try this approach with 38 • 42. First find the midpoint, 40. Then 38 • 42 = $(40 - 2) • (40 + 2) = 40^2 - 2^2$. Voilà! You got the product in your head without a calculator and without paper and pencil.

We have used the fact that $(n - a) • (n + a) = n^2 - a^2$—proven as a consequence of the distributive property to be reduced to the difference between two squares (using both geometric and algebraic demonstrations in chapter 2).

We may have pulled a fast one. First, is calculating squares of numbers so easy? Second, all the multiplication examples so far have had a midpoint that was a natural number. Suppose we wanted to calculate 17 • 22 with the same approach. The midpoint (n) is 19.5, and the distance between the midpoint and each endpoint [$(n - a)$ and $(n + a)$] is 2.5. So to calculate 17 • 22, we need to calculate $(19.5 - 2.5) •$ $(19.5 + 2.5)$. Then we would have to get the following difference of squares:

$$19.5^2 - 2.5^2$$

For positive integers, locating midpoints does not appear to get more complicated than using fractional values that end with ".5." But how do we handle this difficulty? In chapter 2, we explored some mental tricks that we can apply. We showed how to handle the following such numbers:

$$55^2 = 3025$$

$$65^2 = 4225$$

We got the answer in each case by appending 25 to the product of the first digit and that digit's successor. For example, to mentally calculate 65^2, multiply the first digit (6) by that same digit plus 1 (7). Hence we calculate 6×7, which equals 42, and append "25."

What would we do with 19.5^2? If for the moment you ignore the decimal point and consider the desired problem to be 195^2, then a reasonable procedure would be to multiply 19 by its successor, 20. Again, you can do that in your head, yielding 380. Extending the procedure we used for squaring a two-digit number ending in 5, we would have 38,025 as our answer for 195^2. When we consider that we are dealing with decimals, the answer would be 380.25. You can explore this approach by extending the procedure we used for one-digit numbers ending in 5.

Reflections

I have intentionally invited you to follow my thinking that progressed from some empirical observations instead of presenting the "jewel" in a neat manner with clear statements and proofs along the way. It took a while for me to make clear what was going on, and though you may have anticipated where I was headed,

I tried to persist in making this a journey rather than a set of conclusions. It all began with some intrigue over a pattern of squares that was emerging for not one but two sequences of squares (somewhat forced in column 4 but observed as a consequence in column 5) in table 3.1. The data derived from representatives of the three sequences (figs. 3.1, 3.3, and 3.6) suggest that we are headed in an interesting direction.

Table 3.1
Three selected entries from figures 3.1, 3.3, and 3.6

From figure	Factors	Product	Strives to be	Misses by
3.1	5 • 7	35	36	1
3.3	3 • 7	21	25	4
3.6	3 • 9	27	36	9

What eventually emerged was a statement that was clear and worth proving: that we could find the product of two numbers in a new way—and one that depended on a form of the distributive property and that involved square numbers in two ways, as indicated in the last two columns of table 3.1. To put a fine point to it, and focusing on two factors x and y to be multiplied, then

$$x \cdot y = (\text{midpoint})^2 - (\text{distance from midpoint to either endpoint})^2, \qquad (4)$$

which symbolically reduces to

$$x \cdot y = \left(\frac{x+y}{2} \right)^2 - \left(y - \frac{(x+y)}{2} \right)^2 \qquad (5)$$

using the summary in equation (4).

Again the geometric proof of the extended distributive property $[(a + b) \cdot (a - b) = a^2 - b^2]$, as worked through in figures 2.2–5 in chapter 2, is easy to follow, though the algebraic one is a bit tedious. And as shown, the supposed messiness of a fractional midpoint is easily overcome.

More important, however, this procedure raises questions about its relationship to the conventional approach for multiplying positive integers. The new procedure can raise several mathematical and philosophical problems. Philosophically, we would profit from discussing which procedure is simpler and which is more complex. Playing around with this musing in relation to the standard one makes us aware that "simple" and "complex" are context-bound expressions. What is missing when we say that "*A* is simpler than *B*" is really an ellipsis for a missing *Z*, as in "*A* is simpler than *B* with regard to *Z*." For example, if your goal is to calculate quickly without use of instruments, which would be more inviting? Also, because this procedure depends primarily on finding the difference of two squares, what procedures (other than the one for squaring numbers ending in 5) would you create?

What would happen if you expanded this procedure beyond the positive integers? Negative numbers? Fractions? Can you apply some of this thinking in seeking an alternative to long division?

Finally, regardless of ease of calculation, is there not something aesthetically attractive about reducing the problem of multiplication to that of finding the difference of two squares? Might we devise other aesthetically attractive procedures as alternatives to standard algorithms?

Many of these questions have interesting educational implications. One might be that such musing around takes time and competes with the standard fare of "covering" curriculum. What might you consider to be its educational payoff and educational deficits as well? See Brown (1975) for other questions and for further issues from which I derived this chapter. An observation from a nine-year-old inspired another new algorithm useful for division; she originally did not want to pursue her observations because "it's not the one the teacher taught the class. If I show it to her she'll tell me . . . that maybe when I grow up and understand about such numbers, I can do division that way" (Brown 1981, p. 12). What do you see as the pros and cons of that statement from an educational point of view?

References

Brown, Stephen I. "A New Multiplication Algorithm: On the Complexity of Simplicity." *Arithmetic Teacher* 22 (November 1975): 546–54.

———. "Sharon's 'Kye.'" *Mathematics Teaching* 94 (March 1981): 11–17.

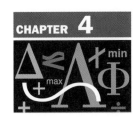

CHAPTER 4

A Show-Stopping Belt

Modern Technology

Modern technology allows us to detect light from stars even years after their demise. Given such advances, we have a modest problem that astronauts orbiting Earth might actually carry out (or at least simulate). And then they could step back some distance to view their handiwork. The procedure would require a large amount of beltlike material and considerable patience. Perhaps astronauts who have a telescope focused on the equator could make these observations.

Or we could just sit in an armchair and analyze it ourselves, with modest knowledge about how radii and circumferences of circles behave. Here is the problem:

> Operating like Superman, place a thin belt around the equator. Then, add 30 feet to it, and place the new belt around the equator so that it is concentric with the original one. What would the distance between the two belts be?

Figure 4.1a is a picture I took of the belt problem when I was recently in outer space (or perhaps just feeling spacey while photographing a manhole). Figure 4.1b is a sketch useful for analyzing the problem.

Analyzing the Problem

If r signifies the radius from Earth's center to the equator, what would the distance (Δr, meaning the change in r) between the two belts in figure 4.1b be?

Imagine that you are in a rocket ship orbiting Earth and you can observe the new belt that has been placed around the old one and concentric with the circumference of the earth. If you were ten miles above Earth, what would the difference look like between the two belts?

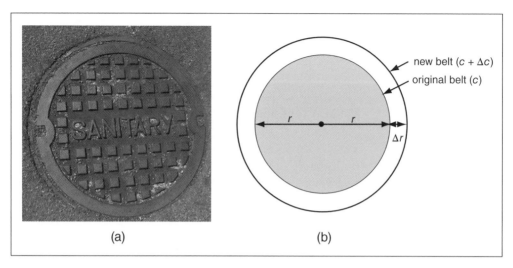

Fig. 4.1 (a) A manhole cover as a two-dimensional cross section of the belt problem. (b) A sketch of the problem.

Come up with a first approximation without using any technical information other than that Earth's circumference is about 25,000 miles and you are adding only 30 feet to the circumference. So we would be slipping Δr between a circumference of 25,000 miles (the original belt) and the new one of 25,000.0057 miles (because a mile is 5280 feet). If you prefer feet rather than miles, their circumferences would be about 132,000,000 feet and 132,000,030 feet, respectively.

What would the distance between the two belts be? Which of the following might fit between them? A period at the end of a sentence? An ant? An anthill? A full-grown aunt? The Empire State Building?

Most people I have asked these questions have come up with an intuitive guess (in the absence of actual calculation) that it would probably be less than an ant (not a fully grown aunt at that).

Let's see what happens with some straightforward calculation. We will use the relationship of the circumference of the circle to the radius: circumference = $2 \cdot \pi \cdot$ radius.

This relationship has a long and glorious history. Various approximations for π in ancient China assumed that it was about 3. It also appears in the Bible measured in cubits. Archimedes was the first mathematician credited with an approximation based on geometric properties. He essentially inscribed and circumscribed regular polygons around a circle and calculated π as the set of polygons converged on the circle. In about 240 B.C., he showed that π was between $223/71$ and $22/7$, or about 3.1416 (Eves 1969).

Others later used various algebraic and trigonometric procedures to better approximate π, but for many years people hoped to find a precise fraction to define the value of π. The search continued until 1767. Although it was then possible to approximate π to many decimal places (which with modern technology must be more than 1 million by now), Johann H. Lambert proved that π was irrational (could not be expressed as a fraction). In our calculation, we will use $^{22}/_7$, or 3.14, as a good approximation of π. To impress friends, however, you might let them know that a greater degree of accuracy (to twenty decimal places) is 3.14159265358979323846. When we are done, you might want to figure out how much more accurate we would have been if we had used the twenty-decimal-place figure for π. To begin the relatively simple algebraic analysis, we will use the following data:

- The circumference of the equator is 25,000 miles.
- The new belt is 30 feet longer.
- Circumference = 2 • π • radius.

Because we already know the circumferences of the original circle and the one that is 30 feet longer (25,000 vs. 25,000.0057 in miles), we can figure out the radius of each circle. To start, we could use the relationship of the circumference to the radius in both cases. So, $25,000 = (2\pi)r$ and $25,000.0057 \approx (2\pi)(r + \Delta r)$. If π ≈ 3.14, then we have 25,000 ≈ $6.28 \times r$ and 25,000.0057 ≈ $6.28 (r + \Delta r)$. Verify that the values of the two radii, $(r + \Delta r)$ and r, are approximately 3,980.8926 and 3,980.8917, respectively. The difference (Δr) between the two radii is therefore approximately 0.0009 miles, or 4.8 feet. How does that compare with what you guessed? Is it closer to the height of the Empire State Building? An ant? An aunt? An anthill?

Distributive Property to the Rescue

The conclusion that the increase in radius is about 4.8 feet when the circumference of the earth is increased by a relatively small amount (30 feet added to 25,000 miles) is astounding. But invoking the distributive property in the analysis yields an even greater surprise. Consider what happens then. In the analysis, we will minimize reference to specific values and see what we arrive at when we do so. Referring to figure 4.1b, if we focus on the large circle, we have this:

$$c + \Delta c = 2 \cdot \pi \cdot (r + \Delta r) \tag{1}$$

If we look at the right-hand side, we smell a rat: the distributive property waiting to be harnessed. Using the distributive property, we get this:

$$c + \Delta c = 2 \cdot \pi \cdot r + 2 \cdot \pi \cdot \Delta r \tag{2}$$

Because c and $2 \cdot \pi \cdot r$ are the same, we subtract it from both sides of (2), to get the following:

$$\Delta c = 2 \cdot \pi \cdot \Delta r \tag{3}$$

Because we are seeking the value of Δr, we divide by $2 \cdot \pi$ and get this:

$$\Delta c/2\pi = \Delta r \qquad (4)$$

We could have introduced the value of Δc earlier in the algebraic analysis but have lost nothing by waiting until the end. So we have the following:

$$30/2\pi \approx 30/6.28 \approx 4.8 \qquad (5)$$

If Δc is 30 feet, then the value of Δr is almost 5 feet. Is this what you guessed?

But something more astounding is in store as we continue to reflect on where we have arrived.

So What?

This analysis leads to a phenomenal conclusion. We began by looking at an approximation for the circumference of the equator (about 25,000 miles) and then compared that with a circumference having the same center but 30 feet longer. We guessed that the distance between the two circumferences might be minuscule. With some calculation, we found that the intuitive guess differed significantly from what the calculation disclosed.

We will return to this disparity shortly, but let's notice something else that has snuck under the rug (or between the belts). Look what the calculation in equation (5) depends on. What happened to the original circumference of the equator? The only variable that influences the distance between the original belt and the new one is the increase of the original belt, not the actual length of the original one. So, 25,000 as an approximation of the circumference of the equator is irrelevant for finding the distance between two concentric circles.

What does that mean? Of course, we might just admit that our intuition was wrong and move on to other matters. But if we feel uncomfortable with the disparity of what is and what we expected, we might do several things. The most obvious would be to see what we can do to bring our intuition into alignment with the "reality." However, if we hold our intuition to be precious and not easily modified, we could try to figure out where the mathematical analysis might be fragile and perhaps in need of modification. What else might we do here?

Some Playful Options

In concluding this chapter, we will examine the potential of the "belt around the earth" problem to generate some playful thought. For a start, if we find the surprising conclusion to be tantalizing, we might ask questions such as these: If the actual circumference is irrelevant for finding the distance between the two circles, then could we start with a circle considerably smaller than the equator and add 30 feet to the circumference? What might you start with? Take a limiting case (such as an original circle with the smallest perimeter you can imagine).

Another source of investigation might be the following: If our find is so surprising with two concentric circles as a starting point, what would happen with other concentric figures? Suppose we began with a square and then created a new square with the same center as the original one but with a greater perimeter. See figure 4.2.

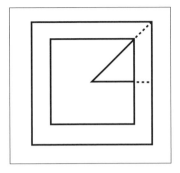

Fig. 4.2. Seeking concentric squares as an analogy

Perhaps try a square with a perimeter that is the same as Earth's circumference—about 25,000 miles—and again add 30 feet to that perimeter. What would be the difference in length between the two squares? The analogy with the two concentric circles breaks down a bit, because (as suggested by the two dotted lines in fig. 4.2) the difference between the two squares seems not to be constant.

How significant is the breakdown of the analogy between the circles and squares? Instead of merely observing the lack of analogy, could we explore other questions?

Select another figure, perhaps a triangle easy to work with, and then surround it with another triangle the same shape as the original.

Why did our intuitions about this problem go so awry? Discuss the issue among colleagues and friends. Where would you try to resolve the difference between your intuition and what you are finding out by calculation? This may not be an easy question to work through. Mathematics has many paradoxes or apparent paradoxes for which intuitive expectations and logical analyses clash. Resolution often begins with an effort to find out possible mistakes in calculation. Sometimes you have to reeducate your intuition. Sometimes the conflict creates a problem, a conceptual block perhaps, that takes a long time to unravel. Sometimes the inability to reconcile the logical and intuitive disparity may discourage further exploration of the problem. Some of these issues will come up again in chapter 7 (e.g., in discussing the advice Papa Bolyai gave to his son regarding the son's effort to pursue studying the early stages of non-Euclidean geometry).

Reference

Eves, Howard. *An Introduction to the History of Mathematics.* New York: Holt, Rinehart, and Winston, 1969.

CHAPTER 5

The Distributive Property: Heading toward Isomorphism

The only man I know who behaves sensibly
is my tailor; he takes my measurements anew
each time he sees me. The rest go on with their
old measurements expecting me to fit them.
— *George Bernard Shaw*

At Grand Central Station, I sit and watch
the mad rush of people coming and going.
The actors change. What catches my interest
momentarily is the similarity of their stories.
What intrigues me for a lifetime is imagining
the differences.
— *Gilbert Brenner*

Preliminary Remarks

The concept of "same/different" is built into nature itself
and most likely has a direct impact on the evolution of a
species. One of the most instinctive reactions of any organism is to determine whether a newly encountered creature
is friend or foe. The well-known "fight or flight" reaction
is predicated on the assumption that an organism is being threatened by another creature that it is different from
itself in some important way.

Some organisms have a built-in imprinting response.
Thus, ducks not only will follow their mother wherever she
goes shortly after they are born but also will do so with
other "organisms" that are similar in important ways to the
mother. Ducks even will be compelled to follow inanimate
objects that resemble mother ducks. Of course, the vital
question here is, what do we mean by "similar"? What at-
tracts ducks to a mother "look-alike" is not shape or size,
as we might predict. Rather, the similarity of the "organism's" pitch to that of the mother duck determines whether
such imprinting will occur. Thus, a baby duck will imprint
on a roboticized tape recorder if the sound on the tape is
the same pitch as the mother's.

Cultures have fine-grained sensitivity to the concept
of similarity and difference. Many outsiders would be

struck by the similarity in looks and culture between Arabs and Jews, Turks and Greeks, North and South Koreans, citizens of Northern Ireland and the Republic of Ireland. In-group members in each case, however, are acutely aware of the differences—so much so that people marshal all sorts of means to intimidate, harass, and even eradicate the other just for being a member of the out-group.

Part of what it means to be tolerant is to perceive these differences and to have them not matter in some important ways. Thus, it is a significant first step for members of an in-group to realize that others may be different but not to discount people because of that difference. The concept of equal protection under the law is to acknowledge that although major differences may exist among people, people nonetheless deserve equal protection because they are human. Part of the role of education in fact may be to discourage us from perceiving some differences as significant. But of course, we would not survive long if no differences were perceived. For example, I must be able to tell the difference between my wife and someone else's wife. A mother must be able to identify her own child.

The concept of "same/different" is also deeply embedded in mathematics. Consider congruent figures. Two triangles are congruent if they are essentially identical, but not necessarily located in the same spot. If they are congruent, however, we can create a set of transformations that "slides" one on top of the other.

Another example would be equivalence classes composed of the integers. From many points of view, the numbers 4, 7, 10, and 13 are quite different. However, they have the same reminder when divided by 3.

The following anecdote captures how two different mathematical structures operate in essentially the same ways. We will explore the issue mathematically after the fanciful dream below. Be patient: our discussion will act as a form of calisthenics for more vigorous activities.

A Dream and Its Interpretation

Shmuel, an elderly man living in the mid-eighteenth century in Europe, is exhausted after milking the cows and slaughtering the chickens. He lies down on a soft bed of straw in the barn. As he lies down, the memory he recalls is the good news he has just heard from his daughter—that she is in her first trimester of pregnancy.

As he nods off, he dreams about his forthcoming grandchild and wonders whether it will be a boy or a girl. If it is a girl, his son-in-law will have to think about how to prepare a proper dowry in about twenty years. Shmuel wonders whether he himself might contribute to that dowry by passing along some of his land, which he imagines will be rich in corn in years to come. If the child is a son, he imagines that the boy will join both his son-in-law and himself in cutting the grass. It crosses his mind that he might also like to have a granddaughter with whom to talk about poems he might like to write when he is too old to handle the farm. He is imagining what family members the child might resemble, what the first words might be, when suddenly he sees a brightly colored horse-drawn carriage and an old woman dressed with emerald beads on her blouse and skirt. She beckons him to step up to the seat next to her on the carriage. He is shy and does not feel comfortable sitting next to a strange woman, but he steps aboard as she whispers that she has mysteries to reveal.

They chat about the weather for a while when suddenly the horse begins to take flight, hovering only a few feet above the ground at first, but eventually Shmuel can just barely see his farm from the air. After a seemingly interminable time, he approaches stars from a distance. He hears voices of his daughter and son-in-law—barely audible but still coming from the direction of his hometown. They are telling him that he is in for an adventure and that he should go with the flow and resist the attempt to make logical sense of what is coming his way.

Shmuel begins to fear that he will never return. He then hears his daughter and son-in-law singing, but their voices are no longer coming from his farm. They seem to be coming from a strange village miles away from his own. The horse stops abruptly and lands on a piece of earth. At first it looks as if he has returned to his own village. He sees the farmhouse in the distance, the plot of land where he will plant corn, and several people coming toward him from the distance. He embraces his wife, his daughter, his son-in-law. They seem to be the same as when he left on his horse-drawn adventure. They have the same clothing and smells, and they talk the same—but something is different. Suddenly he realizes that, though they all have the same features, they are about one-tenth their normal size.

His daughter, whom he is hugging, hits the ground with a thump and he realizes that he too has begun to shrink. His wife has prepared a picnic of his favorite dishes. They toast his joyful return with some wine used only for special occasions. As he always does, Shmuel prepares to tell them about his adventure, when suddenly his wife calls him in from the field for dinner. He is just finishing dinner when she tells him that they are ready to eat breakfast before going to visit her family in the next village.

Shmuel asks Mordecai, the village wise man, to interpret the dream. Mordecai asks Shmuel what he was doing before he began to dream. Shmuel tells him that he was busy trying to see in a clear light why it was that when he calculated the number of eggs his chickens had laid in the past two months, the totals of 470 and 220 could be calculated more easily than he had previously thought. He realized that he could ignore the zeros at the end of each count and, once he had gotten that sum, he could insert a zero at the end. It was an amazingly simple idea, he thought, but he wondered whether there might be more to it than appeared on the surface. After some questioning, Mordecai confirmed that the dream indicated that Shmuel was indeed obsessed with the sex and appearance of his forthcoming grandchild. (This dream appears originally with less elaboration in Brown [2001, pp. 229–31], as part of a mathematical Talmud created for teaching about similarity and difference in mathematics.)

Mordecai calls together some of the elders in the community to figure out what is going on. He includes the village psychologist and the mathematician. After much probing of Shmuel's temperament and his dream, Ruchel, the psychologist, concludes that time was accelerated in the dream because Shmuel was so anxious to find out all about his forthcoming grandchild before his daughter would give birth. Both Ruchel and the mathematician, Izzy, pondered how they might portray what was happening and the relation between the arithmetic of Shmuel's egg counting and the prediction of characteristics of his forthcoming grandchild. First Ruchel drew the following picture in the sand (fig. 5.1):

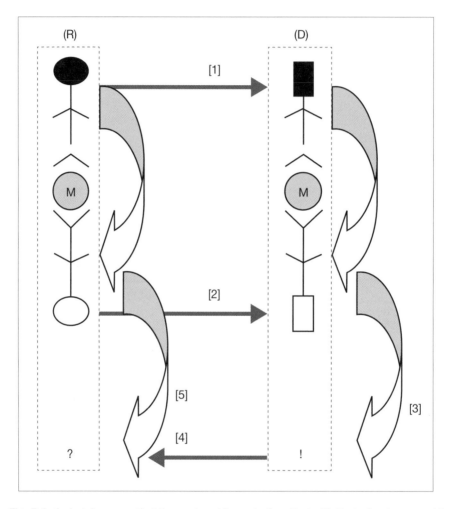

Fig. 5.1. A sketch connecting the real-world event of mating with that of a dreamworld counterpart (reprinted from Brown 2007, p. 20)

Ruchel explained that on the left side of figure 5.1 was R, standing for some real-world event. On the right side was D, standing for a dreamworld. On each side were two stick figures, standing for a man (darkened head) and a woman (clear head). In the real world (R), the heads were circular; in the dream world (D), they were square.

Ruchel explained that the circled M stood for an activity that took place in both the R and D worlds—an act of mating in both systems. "What then are all the arrows and numbers about?" asked Shmuel. Izzy then offered this explanation, which you may already have anticipated:

> [5] [in fig. 5.1] signifies the actual results in the real world of the mating nine months after the event. But that is precisely what Shmuel wants the answer to but cannot get directly. So, he takes an indirect path. He transports the male and female objects from (R) onto the (D) world as indicated in the movements [1] and [2]. Once these objects are in the dream world, they can mate, but since this world moves at a much more rapid pace, the grandchild (indicated by "!") is created almost instanta-neously [with (3) signifying the result of the mating act]. Then all that is left is for Shmuel to transport "!" back to the real world (R) signified by arrow (4). (Brown 2007, p. 19)

That sketch is all that Izzy needed. He then drew a comparable sketch in the sand (fig. 5.2):

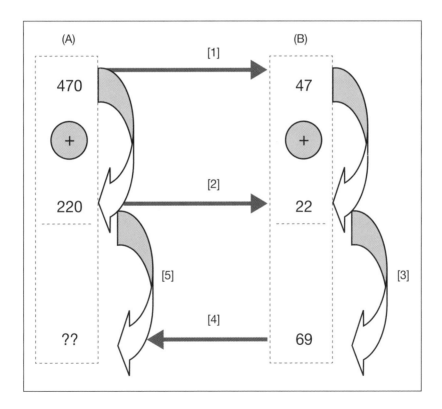

Fig. 5.2. A mathematical shortcut analogous to the real-world act of mating.
(See fig. 5.1 for source.)

The two sketches are comparable. Unlike the prediction of information of his grandchild nine months early, he could have actually calculated the sum of 220 and 470 directly, but if we assume he had not previously seen the shortcut in this light, it might have been an insightful leap. The moves from (A) to (B) and then back to (A) in the arithmetic case were essentially the same as the ones described in the grand-child-prediction case. The "operations" were different (one multiplying numbers and the other "mating") and the elements were different (numbers and people), but the scheme is the same.

For us (as opposed to Shmuel), to observe the shortcut for adding two numbers ending in zero in figure 5.2 is perhaps like a Mack truck pulling a kiddie car. But it is useful for understanding how the distributive property is a hidden culprit that justifies the movement. More precisely, the distributive property can itself be thought of as a form of mapping from a system into itself.

In this setup, we have two systems, (A) and (B), each with elements (such as numbers) and operations (ways of joining the elements in each system). Then we need some sort of "mapping" that associates elements in one system with those in the other. The association is helpful when the operation performed in one system is simpler than that in the other, so that we can map the results of the second system back into the first.

This is a quick and oversimplified introduction to a concept, which when more fully elaborated can depict an "isomorphic-like" structure. To be accurate, it needs additional restrictions and a more careful definition. Anyone interested in a more precise definition should consult a text on modern algebra, such as Judson (1994), a free Internet download. Brown (2001, pp. 117–25) includes a more elaborately developed intuitive extension of this section.

Here, the "mapping" ([1] and [2]) associates elements from system (A) with those in system (B), via the mapping $n \rightarrow (1/_{10}) \cdot x$ for each element in (A), and then sends the results (47 + 22) back to system (A) by the reverse mapping from (B): $y \rightarrow (10) \cdot y$. (At this point, many of the elders began to doze off, having misconstrued "mapping" as "napping.")

You may of course wonder what happened to the distributive property in this analysis. In going from (A) to (B), you are multiplying each of the original numbers by $1/_{10}$, thus yielding $1/_{10}$ of 220 and $1/_{10}$ of 470. You then have added their sum (47 + 22) in (B); before sending that sum back to (A), what did you multiply that sum by? See whether you can explicitly state how the distributive property has been implicated.

More than a Bump on a Log

Logarithms are a bit more enticing as a labor-saving device (or at least were so before computers) but offer a similar kind of calisthenics. Though calculators make using tables of logarithms obsolete (though the concept is particularly relevant from a more theoretical point of view in the study of calculus), it is of conceptual interest because the operations in the sets we establish for (A) and (B) are not the same.

Without going into basic definitions, we can use the scheme of figure 5.3 to transform multiplication problems into addition problems by using logarithms. Here is a brief sketch:

For the last step, we can find $\log n + \log m$ and locate the number associated with that log. Because $(\log n + \log m) = \log (n \cdot m)$, we can select this backward step for determining $n \cdot m$ (or at least for coming up with a good approximation).

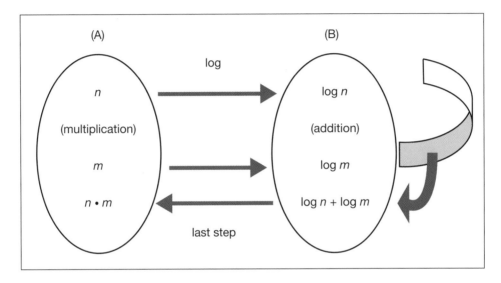

Fig. 5.3. Calculating logarithms as a form of mapping between two systems

Beyond Calisthenics: Isomorphic Structures

Despite the lack of precision, we have used the concept of isomorphism-like structures to talk about an issue we raised at the beginning of this section. That is, we wanted to explore two structures that have a fundamental sameness even if they are significantly different on the surface. That sameness of structure is often useful if it is easier to work in one structure than the other. That is, the elements of the two systems may be different and the operations in the two systems may be different, but nevertheless the two structures (consisting of elements and operations) act as if they are the same. For the relationship between the two systems to actually be isomorphic, a one-to-one correspondence must exist between the elements of the two systems. To present a simplified version of the structure for readers unfamiliar with some terms not defined explicitly (such as *mapping*), I do not specify the numbers in each domain. Part (A) of figure 5.2, for example, purposely focuses on natural numbers that are multiples of ten.

To see how the concept functions in more accurate and precise terms when both the operations and the elements in system (A) and (B) are different, we will look at two new systems: one a simple clock with just three numbers, 0, 1, 2, and the other a geometric scheme that uses the equilateral triangle. Figure 5.4 represents a simple clock.

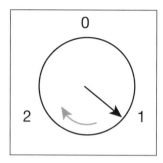

Fig. 5.4. A three-digit clock

Table 5.1 shows an addition table for adding hourly increases.

Table 5.1
Addition table for adding a three-digit clock

+	0	1	2	(Time)
0	0	1	2	
1	1	2	0	
2	2	0	1	

For the second structure, we look at an equilateral triangle with a pinhole in its center (fig. 5.5).

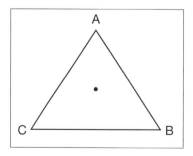

Fig. 5.5. An equilateral triangle with a pinhole in the center

If you "spin" it clockwise 120 degrees around the small pinhole, you end up with position of $\triangle CAB$—the one to the left in figure 5.6. If you spin it another 120 degrees clockwise, you end up with the position of $\triangle BCA$—the one to the right in figure 5.6.

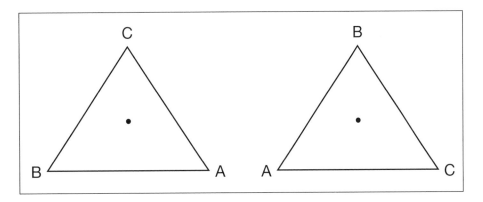

Fig. 5.6. Spinning an equilateral triangle 120 degrees around the center

So start with the triangle in the position of figure 5.5. Call that position X. If you rotate that triangle 120 degrees clockwise, you end up with the triangle on the left side of figure 5.6. Call that new position Y. Repeat the same rotation and you have position Z—which would be the same as if you had rotated figure 5.5's triangle by 240 degrees clockwise.

See what happens if you start with any of the other triangles and use the same rotation scheme: two rotations of 120 degrees, labeling each iteration X, Y, and Z. Now start with any of the three triangles and choose any two successive rotations (from the degree measures we have used). Designate these rotations as R_{120}, R_{240}, and R_{360}.

You can now look at the rotations themselves as the elements of the system. We can depict the operation (combining these elements) with an asterisk, which means "followed by":

$$R_{120} * R_{240} = R_{360}$$

Table 5.2 shows this information by locating the first element, R_{120}, in the column on the left (the first deed performed) and by locating R_{240} in the row on the top. The box that describes the result, R_{360}, is at their intersection.

Table 5.2
Summarizing three equal spins of an equilateral triangle

*	R_{120}	R_{240}	R_{360}
R_{120}	R_{240}	R_{360}	R_{120}
R_{240}	R_{360}	R_{120}	R_{240}
R_{360}	R_{120}	R_{240}	R_{360}

From the viewpoint of sameness and difference: though tables 5.1 and 5.2 have differences (the elements being hours in table 5.1, whereas they are rotations in table 5.2, and the operation being a form of addition in table 5.1, whereas it is * in table 5.2), the two structures are identical. They are different in composition but operate the same way.

Elements in the two systems exhibit a one-to-one correspondence: in clock arithmetic, 0 corresponds with R_{360} in the triangle rotation; 1 corresponds with R_{120}, and 2 corresponds with R_{240}.

Clock arithmetic is essentially the same as modular arithmetic. The set {0, 3, 6, 9, . . . 3n} can be thought of as 0; {1, 4, 7, 10, . . . 3n + 1} can be thought of as 1; and {2, 5, 8, 11, . . . 3n} can be thought of as 2.

In retrospect, reexamine Shmuel's search for a prediction of the sex of his grandchild in relation to the addition of numbers ending in 0. The connection between the rotation of equilateral triangles and the three-digit clock leads to a sharper conception of isomorphism than that of numbers ending in 0. We began with only three-digit numbers ending in 0 depicted in figure 5.2. What would have been the case had we chosen any two numbers (regardless of the number of digits) ending in 0? Second, we focused on associating movement from (A) to (B) in figure 5.2. Suppose we wanted to move in the other direction: how would we specify the numbers in (B)? Furthermore, specifying the one-to-one correspondence in the two finite sets is easy. What could it have been for adding numbers ending in 0 for set (A) and those in set (B)? Playing around with various "what if nots" should help enrich and clarify the difference between two systems and more precisely convey the concept of isomorphism and of isomorphism-like structures.

References

Brown, Stephen I. *Reconstructing School Mathematics: Problems with Problems and the Real World*. New York: Peter Lang, 2001.

————. "Transcending the Kpelle Nightmare: Personal Evolution and Excavations," *Philosophy of Mathematics Education Journal* 22 (November 2007). http://people.exeter.ac.uk /PErnest/pome22/.

Judson, Tom. *Abstract Algebra: Theory and Applications.* Boston: PWS Publishing, 1994. http:// abstract.ups.edu/download.html.

CHAPTER 6

A Field Day

Playing Fields and Meta-questions

Students who have studied the axiomatic structure of number systems are aware of various formal properties of systems, such as the natural numbers, integers, rational numbers, real numbers, and imaginary numbers. Students have come across the commutative and associative property, the property of closure, the existence of identity elements, and of course the distributive property. People who have learned about systems with these properties may believe that these were always recognized as number properties. In fact, the properties were not explicitly explored until the nineteenth century. Eves (1969) summarizes some of the players and their roles.

In the early nineteenth century, algebra was considered simply generalized arithmetic. Instead of working with specific numbers, as we do in arithmetic, in algebra we use letters to represent arbitrary numbers. The early nineteenth-century British school of algebraists—consisting of George Peacock (1791–1858), Duncan Farquharson Gregory (1813–1844), Augustus De Morgan (1806–1871), and others—first noticed the presence of structure in algebra, such as the commutative and associative laws of addition and multiplication, and the distributive law of multiplication over addition.

Back then, the existence of a consistent algebra with a structure contrary to that of the common algebra of arithmetic was inconceivable. But in 1843, Irish mathematician William Rowan Hamilton (1805–1865) invented his quaternionic algebra, in which the commutative law of multiplication does not hold. The next year, German mathematician Hermann Grassmann (1809–1877) published the first edition of his remarkable *Ausdehnungslehre*, which developed whole classes of algebras with a structure different from that of the familiar algebra of arithmetic (Eves 1969, p. 365ff.).

Much mathematical thinking focuses on proving or disproving statements in a given system. Here we will examine questions that use basic axioms, prior proofs, undefined terms, and properties of logic but that have a quite different purpose. We will raise the ante by asking and analyzing metamathematical questions. These are questions less concerned with actually proving or disproving a proposition than with analyzing questions of a more abstract nature—questions concerned with the nature of provability itself.

Perhaps an analogy will help. The United States has a Constitution that spells out its purpose and lays out the laws that define how the various governing bodies will be formed, how they operate, and how they will relate to each other. Some meta-Constitutional questions might be as follows: What does it mean for something to be a law? What are the logical requirements for something to be a law? How specific are the laws? How precise are they? Do any parts potentially conflict with others? Do governance issues exist that the Constitution cannot resolve?

From a mathematical point of view, and particularly with regard to axiomatics and proof, here are some issues we might explore:

- How do the elements of a system relate to each other?

- Is this system complete, in that all propositions that make sense can be proven true or false?

- Can we know whether this system can lead to contradictions?

- Can we know, before actually trying to prove an alleged theorem, that the proof might need specific axioms?

- Does the system include all necessary specific pieces of the basic structure? That is, are parts missing?

Addition	Multiplication
I For any a and b, $a + b$ is a number (closure for addition).	I′ For any a and b, $a \cdot b$ is a number (closure for multiplication).
II For any a, b, and c, $(a + b) + c = a + (b + c)$ (associative principle for addition).	II′ For any a, b, and c, $(a \cdot b) \cdot c = a \cdot (b \cdot c)$ (associative principle for multiplication).
III There is a number, 0, with the property that for all a, $a + 0 = 0 + a = a$ (additive identity).	III′ There is a number, $1 \neq 0$, with the property that $a \cdot 1 = 1 \cdot a = a$ (multiplicative identity).
IV For each a, there is a number $(-a)$ with the property that $a + (-a) = (-a) + a$ (additive inverse).	IV′ For each $a \neq 0$, there is a number $(1/a)$ with the property that $a \cdot (1/a) = 1$ (multiplicative inverse).
V For each a and b, $b + a = a + b$ (commutative principle for addition).	V′ For each a and b, $b \cdot a = a \cdot b$ (commutative principle for multiplication).

And now, the centerpiece that links addition with multiplication:
VI. For all a, b, and c, $a \cdot (b + c) = (a \cdot b) + (a \cdot c)$ (distributive principle).

Fig. 6.1. The eleven axioms for a field (adapted from Brown 1966)

We will focus on one part of an axiomatic system we have been using in much of this book: a field. Chapter 1 mentioned several axioms in the section titled "The Beginnings of a 'Field' Day," and we were drawn to one important element of a field: the distributive property. Figure 6.1 establishes the axiomatic structure for a field with greater focus and precision. Though we have used standard addition and multiplication symbols for the two operations, and 0 and 1 as symbols, those who have studied fields might have seen the description of a field with symbols that do not automatically associate them with familiar number systems. We use the specific association, however, for those who have not previously studied abstract algebraic systems.

What Do You Notice?

As you look at figure 6.1, what do you notice? What questions occur to you? The figure lists eleven axioms, but what more can you say? How do the axioms compare? Do the similarities and differences inspire hunches or observations? Several years ago, these questions led me some metamathematical finds depicted here. My first observations and hunches were less organized. Although I have tried to put them together more coherently, I have also tried to preserve some of the excitement of not quite knowing where and how it would all fit together. Try to put the various pieces of this chapter together on your own as you read through it, to anticipate the metaquestions that are eventually explicit as well as some that might be only implied.

First, though, test your intuition about fields. Look at figure 6.2 and fill in the yes–no boxes. Hold your guesses for now.

		Y	N
1	For all a: $-(-a) = a$.		
2	For all a, b: $(a + b)^2 = a^2 + (2ab) + b^2$.		
3	$1 \cdot 0 = 0$.		
4	For all a, b: $-(a + b) = (-a) + (-b)$.		
5	For all $a, c, b \neq 0, d \neq 0$: $(a/b) + (c/d) = (a \cdot d + b \cdot c)/(b \cdot d)$.		
6	For all a, b: $(-a) \cdot (-b) = a \cdot b$.		
7	For all a, b: $(-a) \cdot (b) = -(a \cdot b)$.		
8	For all $a \neq 0$: $1/(1/a) = a$.		
9	For all $a, d, b \neq 0, c \neq 0$: $(a/b) \cdot (c/d) = (a \cdot c)/(b \cdot d)$.		

Fig. 6.2. Is the distributive property needed to prove these field theorems?

Some Fieldwork

Before thinking further about figure 6.2, what can you observe about the following four theorems based on figure 6.1 before you even prove anything? How do they relate to each other?

(0) For all x: $(-1) \cdot x = (-x)$.

(0′) For all x, y: if $x + y = 0$, then $-x = y$.

(1) For all a, b: $-(a + b) = (-a) + (-b)$.

(1′) For all $a \neq 0$, $b \neq 0$: $1/(a \cdot b) = (1/a) \cdot (1/b)$.

A Proof Relying on the Distributive Property

Suppose you had already proven one or two of the above theorems. Can you use that information to establish any others without starting from scratch? Many of my students have seen the most obvious connection being between (0) and (1). Why? If $(a + b) = x$, then substituting $(a + b)$ for x in (0), we proceed as follows [you may find it easier to see if the two sides of theorem (0) are reversed]:

$$-(a + b) = (-1) \cdot (a + b) = (-1) \cdot (a) + (-1) \cdot b = (-a) + (-b)$$

Which major axiom did we use to establish (1)? Without thinking, just knowing the centrality of you-know-what, what would probably be a good guess? We used a form of the distributive property. Though we used that property, could we have proven (1) without it?

Cheating the Distributive Property

Choosing (1) as a centerpiece, we have already found a relationship between it and (0). We used (0) to prove (1), using the distributive property to do so. In an interesting sense, we may have used the distributive property twice [because we have not yet discussed whether the proof of (0) itself required that property]. Having settled the question of whether the proof of (1) requires the distributive property in its proof, we will then investigate the relationship of (1) to (1′).

A Proof without the Distributive Property

Proving (1) assuming (0′):

How to connect them? If $(a + b) = x$ and $(-a) + (-b) = y$, then we can show that $[(a +b)] + [(-a) + (-b)] = 0$, and thus $-(a + b) = (-a) + (-b)$. It requires use of the axioms on the left side of figure 6.1 to simplify $[(a + b)] + [(-a) + (-b)]$ to yield 0.

Below is a detailed proof of (0′) itself to indicate that (0′) does not require the distributive property for its proof, as well as to show the careful use of axioms from figure 6.1 for those who may not have explored the field properties beforehand.

> We want to show that if $x + y = 0$, then $y = -x$.
>
> If $x + y = 0$, then by axioms I and IV in figure 6.1, $(-x) + (x + y) = (-x) + 0$.
>
> By axiom II, $(-x) + (x + y) = [(-x) + (x)] + y = (-x)$.
>
> By axiom III, $0 + y = (-x)$ and $y = -x$.

Having related (1) to (0) and to (0′), we now compare (1) with (1′). The present we receive is not a proof of (1′) in quite the same way, but one that works neatly by analogy. That is, how do (1) and (1′) compare as statements? If you tried to prove (1′) and had not established the validity of (1), what would you look for as a statement to easily prove it in the same way that we used (0′) to establish (1)?

Look back at the statement and proof of (0′): $x + y = 0$, then $y = -x$. The only axioms we used to verify the statement were ones from the left side of figure 6.1. There 0 is defined as a number that does not change the sum when added to any other number. What do we have on the right side of figure 6.1 that is comparable to 0 for multiplication? If 0 is a neutral element (has no effect on a number it is added to) for addition, what number acts like it under multiplication? The other properties for addition and multiplication (commutative and associative) act analogously as well. The following scheme thus lets us create a new statement (a dual) from an existing one.

Establishing Duals

Given any field statement A, we can form another statement A' (its dual) with the following three principles:

1. Replace addition in A with multiplication in A', and replace multiplication in A with addition in A'.

2. Replace each additive inverse in A with a multiplicative inverse in A', and replace each multiplicative inverse in A with an additive inverse in A'. Here of course we must exclude multiplicative inverses for all 0 elements.

3. Replace 1 in A with 0 in A', and 0 in A with 1 in A'.

Many of the baby theorems we explore in early stages of proving theorems in a field are twins or duals that not only can be stated as twins (using principles 1–3 above) but also can be proven in an analogous manner.

What does it take to see (1) and (1′) as duals of each other? Take the following proposition: if $x + y = 0$, then $y = (-x)$. What do we see as the correspondence that creates duals?

Addition is replaced with multiplication; x and y are the same, but 0 is associated with 1, and $(-x)$ is associated with what on the right side? The answer is $1/x$: x and y stay the same in both; $0 \to 1$, and $(-x) \to (1/x)$. We now have the following analogous statement:

$$\text{If } x \cdot y = 1, \text{ then } y = 1/x.$$

Try to prove it, to verify that it mirrors the proof of (1) in "Cheating the Distributive Property." What would the dual of $(0')$ be, using the correspondence for duals that we have established? For (2) and (3), state their duals and look at the proofs of each of them.

(2) For all a, c, $b \ne 0$, $d \ne 0$, $a/b \cdot c/d = (a \cdot c)/(b \cdot d)$.

(3) For all a, $-(-a) = a$.

Disparity of Truth Values of Duals

Consider the following field statement (with the understanding that x/y is a shortcut for $x \cdot 1/y$, where $1/y$ is defined as in fig. 6.1):

(4) For all a, c, $b \ne 0$, $d \ne 0$, $\dfrac{a}{b} + \dfrac{c}{d} = \dfrac{a \times d + b \times c}{b \times d}$,

essentially the "rule" for adding fractions.

The dual of this statement would be

$(4')$ $(a - b) \cdot (c - d) = [(a + d) \cdot (b + c)] - (b + d)$.

Try to prove (4) by using the axioms of figure 6.1. Show that (4) is not a bona fide theorem by plugging in a few numbers at random. After proving (4), come up with a hypothesis for why a disparity exists in the truth value of these duals.

On the Meaning of "Necessary" in Using the Distributive Property

Do you think that you can you prove (4) without the distributive property, just as we showed we could prove that $-(a + b) = (-a) + (-b)$ both with and without the property? You probably used the distributive property (or some derivative of it) in proving (4), so let's reexamine your answers to figure 6.2: whether the distributive property is necessary in the proof.

Even before coming up with a proof using the distributive property, I suggested a scheme that would tell us that the distributive property is not necessary to prove (1). Look again at $(1')$, the dual of (1):

$(1')$ For all $a \ne 0$, $b \ne 0$: $1/(a \cdot b) = (1/a) \cdot (1/b)$.

That is, knowing that the original statement and its dual are both true suggests that the distributive property is not required to prove either. Why?

A Fascinating Side Path

Having compared (4) with (4′), we indicated that the disparity of truth value of duals is enough to suggest that the true one requires the distributive property for its proof. But perhaps we have gone overboard in seeking duality of statements as a way to determine whether a proof needs the distributive property. Consider proposition (5) and its dual:

(5) For all a, $a \cdot 0 = 0$.

(5′) For all a, $a + 1 = 1$.

We can prove (5) with the distributive property. Doing so is simple, though it begins with a tricky, nonintuitive move: $0 = 0 + 0$—not false, but a bit unexpected perhaps. Then $a \cdot 0 = a \cdot (0 + 0)$. Where to from here? The distributive property, of course. So $a \cdot 0 = a \cdot (0 + 0) = a \cdot 0 + a \cdot 0$. But if $x = x + x$, it is a small step to prove that x must equal 0, and so $a \cdot 0 = 0$.

Now before we even look toward its dual (5′), this example opens up another direction. It looks as if (5) uses only one operation, multiplication. If so, then its proof should be encapsulated entirely in the right side of figure 6.1, and therefore we should be able to prove (5) without the distributive property.

This is a delightful observation. Think about it before comparing it with its dual. Look carefully at the full statement of (5). Does it really portray only one operation, multiplication?

What elements are in the statement itself? In addition to using multiplication, it uses 0. What operation is 0 associated with? Aha! In a disguised way, the statement of this theorem embodies two operations.

One reason to compare a statement and its dual, other than just focusing on the number of operations in a theorem, is that how many operations are involved may not be apparent. Even if two operations were in play, however, you can imagine a statement that has (let's say) two operations explicitly involved and yet does not require the distributive property in its proof.

Try $(a \cdot x) + b \cdot c \cdot d = d \cdot b \cdot c \cdot (x \cdot a)$. It is trivial, perhaps, but whether a proposition is trivial in the sense that this one is may not be apparent. So, back to the drawing boards: (5) and (5′) are duals of each other, but with different truth values. Thus, (5) does require the distributive property in its proof.

A Leap of Imagination

Let's loosen up a bit. Let's try to prove (5′), even though we know that it is false that $a + 1 = 1$ (another statement with two operations in disguise). Following is a proof of (5′) that mimics a proof of (5) but that uses dual replacement properties I–III in figure 6.1, disregarding that the statement itself is false:

$$a + 1 = a + (1 \cdot 1) \text{ [Instead of } a \cdot 0 = a \cdot (0 + 0)]$$
$$(a + 1) = a + (1 \cdot 1) = (a + 1) \cdot (a + 1)$$

Wait. How did we get $a + (1 \cdot 1) = (a + 1) \cdot (a + 1)$? We have created a feat of legerdemain. We used the dual of VI′—an attempted dual of the distributive property—not included in figure 6.1. That is, instead of using $a \cdot (b + c) = a \cdot b + a \cdot c$, we sought the very thing that seems to destroy the symmetry of figure 6.1. We have created the dual of the distributive property itself. Using I–III, we have VI′: $a + (b \cdot c) = (a + b) \cdot (a + c)$. Applying the leap of faith with (5′), we have $(a + 1) = (a + 1) \cdot (a + 1)$, which means that $a + 1 = 1$ or $a + 1 = 0$.

$$(6) \quad \text{For all } a \neq -1, \, a + 1 = 1.$$

Where Does This Lead Us?

We got the result we did by asserting (frivolously?) not only that axioms I–V in figure 6.1 had analogues in the form of axioms I′–V′ but also that VI, the distributive property, has a dual that is also an axiom. Actually, the dual of the distributive property (VI′) does hold for some elements in the set of real numbers, but not for all elements of the set. For a side journey, you may wish to figure out when it does hold in that set. Try expanding (VI′):

$$a + (b \cdot c) = (a + b) \cdot (a + c) = a^2 + a \cdot b + a \cdot c + b \cdot c$$

S.I	For every 2 subsets a and b, $a \cup b$ is a subset of E (closure for union).	S.I′	For every 2 subsets a and b, $a \cap b$ is a subset of E (closure for intersection).
S.II	For every 3 subsets a, b, and c, $a \cup (b \cup c) = (a \cup b) \cup c$ (associative principle for union).	S.II′	For every 3 subsets a, b, and c, $(a \cap b) \cap c = a \cap (b \cap c)$ (associative principle for intersection).
S.III	There exists an element $\emptyset$ (the empty set) having the property that for all a belonging to E, $a \cup \emptyset = \emptyset \cup a = a$ (identity element for union).	S.III′	There exists an element R (the universal set) having the property that for all a belonging to E, $a \cap R = R \cap a = a$ (identity element for intersection).
S.IV	For each element a, there exists an element $R - a$ (the complement of a) having the property that $a \cup (R - a) = R$ (inverse for union).	S.IV′	For each element a, there exists an element $R - a$ (the complement of a) having the property that $a \cap (R - a) = \emptyset$ (inverse for intersection).
S.V	For each a and b, $a \cup b = b \cup a$ (commutative principle for union).	S.V′	For each a and b, $a \cap b = b \cap a$ (commutative principle for intersection).
S.VI	For each a, b, and c belonging to E, $a \cap (b \cup c) = (a \cap b) \cup (a \cap c)$ (distribution of intersection over union).	S.VI′	For each a, b, and c belonging to E, $a \cup (b \cap c) = (a \cup b) \cap (a \cup c)$ (distribution of intersection over union).

Fig. 6.3. Axioms for a Boolean algebra (adapted from Brown 1966)

So: $a = a^2 + a \cdot b + a \cdot c = a \cdot (a + b + c)$. Then either $a + b + c = 1$, or $a = 0$.

So the dual of the distributive property would hold, for example, for $a = \frac{1}{12}$, $b = \frac{2}{3}$, $c = \frac{1}{4}$. Who would have guessed that? Raising the possibility of the existence of the dual of the distributive property has fascinating implications. We might wonder whether a structure exists—other than the set of numbers we have been exploring—for which everything has a dual that holds for all elements in the set, not just those for which $a + b + c = 1$ or $a = 0$.

The algebra of sets (Boolean algebra) is such a structure. Suppose R is an arbitrary nonempty set, and let E be the set of all subsets of R (including the empty set, Ø). Then the system consists of E (the elements of the structure) and the two operations union ($\cup$) and intersection ($\cap$). Venn diagrams offer a model for figure 6.3.

What is interesting about this system (in light of our work with the field of rational numbers) is that because duality functions for all the axioms in Boolean algebra—and in particular the distributive property—we lose the heuristic we have been developing for the metamathematical question we have been posing for the set of rational numbers. Create a statement that is a bona fide theorem in the scheme of Venn diagrams. Look at the dual of that statement and see what you can find. Does using the distributive property in a proof affect the dual of the statement? This would be a fun activity to engage in with a friend who perhaps has played around with Venn diagrams before. For an example, start with three elements for a Venn diagram. Each element will be a set with several numbers. Suppose we select the three sets as follows:

$$A = \{1, 2, 3, 4\}; B = \{2, 3, 5, 7\}; C = \{3, 4, 5, 7\}.$$
$$X \cup Y \text{ consists of all elements in } X \text{ or } Y \text{ (or both).}$$
$$X \cap Y \text{ consists of all elements in } X \text{ and } Y.$$

Each distributive property would be expressed as follows:

$$X \cup (Y \cap Z) = (X \cup Y) \cap (X \cup Z)$$
$$X \cap (Y \cup Z) = (X \cap Y) \cup (X \cap Z)$$

Check out these dual distributive properties for these three sets. If you are visually inclined, you might draw three intersecting circles and put the numbers in place so that common numbers are placed at the intersections of circles (composed of two or three intersections).

Afterthoughts

This has been a chock-full but brief chapter, especially if you have not thought about the basic field axioms that underlie much of arithmetic. The magnet of this chapter has been not only to seek the special feature of the distributive property among the other ten field axioms but also to use it to ask a kind of question quite different from most other statements and proofs in this book. We have not only used the basic components of the structure in stating, proving, and asking questions and making conjectures, but we have also taken an interesting shift of focus: investigating what this system is about and what it can do. The basic question we have explored is well featured in figure 6.2. Reexamine your answers to figure 6.2 in light of reading the rest of the chapter. It may be possible now for you to come up with a different set of answers. Perhaps you can add other questions of the sort that appear in figure 6.2.

Comparing a theorem with the truth of its dual apparently tells us whether we can prove a theorem in a field without the distributive property. That strategy may require some modification, however, because one of the (nondistributive) axioms has a slight nonparallelism to it. Reread figure 6.2 and locate the slightly defective point. Can you tighten the heuristic for determining whether the distributive property is necessary to prove a field theorem or perhaps find other ways to investigate the metamathematical question? Such inquiry forces us not merely to prove things but also to find the significance of a topic we investigate. Edwin Moïse (1965) raises the importance of this issue: "A distinguished algebraist once served as an examiner in a final oral for the doctorate, based on a dissertation on Banach algebras. Toward the end of the examination, the algebraist asked the student to describe some examples of Banach algebras. The student was able . . . to name one example, but his one example was trivial" (p. 410).

As Moïse points out, though the dissertation director could have justified the student's research, the student—who had solved many mathematical problems— had no good intellectual reason to work on those problems in the first place. Do you find that a significant insight regarding the teaching and learning of mathematics in general?

Moïse's concern about significance foreshadows a story about his own life and an event that turned out to be more painful than we normally associate with mathematical thinking—something we will get to in chapter 8.

References

Brown, Stephen I. "Multiplication, Addition, and Duality." *Mathematics Teacher* 59 (October 1966): 543–50, 591.

Eves, Howard. *An Introduction to the History of Mathematics.* New York: Holt, Rinehart, and Winston, 1969.

Moïse, Edwin. "Activity and Motivation in Mathematics." *American Mathematical Monthly* 72 (April 1965): 407–12.

CHAPTER 7

Extending, Expending, and Expanding

The method of postulating what we want has many advantages; they are the same as the advantages of theft over honest toil.

—Bertrand Russell

This chapter will both draw on earlier explorations and expand on questions regarding mathematical and educational thinking. Keep in mind Eves's discussion (chapter 6) of the recent growth of the axiomatic method. Also keep in mind the names given to the various expansions of the number system—from the natural numbers to the negative numbers to the rational numbers to the irrational to the complex—as a signal that regardless of strategies used to go beyond the accepted systems, such efforts often met resistance. In addition to historical questions, this chapter introduces philosophical ones with significant educational implications.

In addition to expanding the behavior of sets that extend the domain of counting numbers, we will look inside that set and examine interesting properties of a subset of the natural numbers: primes. We will gain some better understanding of a property we usually take for granted: reducing fractions to lowest terms.

Finally, this chapter focuses on probably the most famous theorem of geometry: the Pythagorean theorem. We will replace the concept of extension (from geometry to algebra) with one of reversal. We look not at how geometry was expanded to algebra, thus easing the burden of geometric proof, but rather seek greater depth in a geometric conception of the terrain.

Imaginary Numbers

Extending the imaginary numbers met considerable resistance. The source of that resistance is not hard to imagine. Not only did no real-world example or simple model of such numbers exist when people started to think about them, but imaginary numbers also appeared to conflict with accepted tenets. Consider the number i ($\sqrt{-1}$). How does it compare in size with real numbers? Asking whether it is greater or smaller than 0 seems like a reasonable start.

The definition of i as that number that when squared yields -1, however, does not help solve but rather further entrenches the problem. In the set of real numbers, if x is a real number then x^2 must be 0 or positive. It cannot be -1. Regardless of whether $x = -1$ or $x = 1$, x^2 cannot equal -1.

Thus, if we want to incorporate the imaginary numbers within the existing familiar system, we must relinquish the property of order. Without going through the details at this point, think about how you would arrange imaginary numbers on the standard number line. Unlike what happens when you extend from the positive integers to rational numbers, you cannot insert imaginary numbers along an existing number line. More generally, we have to lose a gluttonous disposition built into much of our way of life: seeking more and more without eliminating precious parts of the old way.

Extending number systems, then, is by no means merely a logical act. In any act of extension, we must confront an issue at the intersection of logic and aesthetics. In realizing the limitations of the machinery we have created up to the extension point, we must do more than forge ahead and create something that was nonexistent or meaningless at the time.

Although we must give up some cherished properties in extending number systems, there is nothing God-given about what in particular we must relinquish. Historically, considerable debate, risk-taking, and confusion took place within the mathematical community when deciding what to keep and what to toss if the system was to be maintained as a number system at all.

Multiplying the Integers

Interesting debate surrounds the fact that multiplication works as it does for positive and negative integers, even before we apply its rules to extending imaginary numbers. One issue is framed as pedagogical, but peeling away layers makes it clear that the pedagogy was hiding the important issue of what the distributive scheme was preserving.

One argument (supposedly particularly appropriate for students not too well versed in structural issues) for why the product of two negatives is a positive is based on the completion of a pattern and is presented in a way that does not depend explicitly on axioms. Figures 7.1 and 7.2 are presented in specific cases to determine why the products of negatives and positives should work as they do:

2	•	(3)	= 6
2	•	(2)	= 4
2	•	(1)	= 2
2	•	(0)	= 0
2	•	(−1)	= ?
2	•	(−2)	= ?
2	•	(−3)	= ?

(−2)	•	(2)	= −4
(−2)	•	(1)	= −2
(−2)	•	(0)	= 0
(−2)	•	(−1)	= ?
(−2)	•	(−2)	= ?
(−2)	•	(−3)	= ?

Fig. 7.1. Extending a pattern for multiplication by a positive number $(−2) \cdot (3) = −6$

Fig. 7.2. Extending a pattern for multiplication by a negative number

Figure 7.1 assumes that we already know how to add integers and how multiplication works for nonnegatives. From the pattern, we can determine the last three entries as −2, −4, and −6. Figure 7.2 then adopts the scheme for multiplying a negative by a positive (assuming commutativity for multiplication). From the pattern, we can then conclude that the three bottom entries are 2, 4, and 6. In figure 7.1 we assume that each entry will be two less than the previous line; in figure 7.2, two more than the previous entry.

A supposedly different scheme is based on an axiomatic point of view. The structure usually assumed before the attempt to extend the number system to include the behavior of negative numbers would have some of the field properties for nonnegative numbers (fig. 7.3):

$$a \cdot b = b \cdot a$$
$$(a \cdot b) \cdot c = a \cdot (b \cdot c)$$
$$a \cdot 1 = 1$$
$$a \cdot 0 = 0$$
$$a \cdot (b + c) = a \cdot b + a \cdot c$$

Fig. 7.3. Presumed axiomatic properties of negative numbers

Figure 7.3 summarizes part of the structure for a field discussed in chapter 6. If we know how integers behave under addition (defined as additive inverses) but do not know yet how to incorporate multiplication of negative numbers, we have the following schemes to justify the value of $2 \cdot (−3)$ in figure 7.4 and then $(−2) \cdot (−3)$ in figure 7.5:

$$0 = 2 \cdot (0)$$
$$= 2 \cdot (3 + -3)$$
$$= 2 \cdot (3) + 2 \cdot (-3)$$
$$= 6 + [2 \cdot (-3)]$$

Therefore, $2 \cdot (-3) = -6$.

Fig. 7.4. Product of a positive and a negative by use of distributive property

$$0 = (-2) \cdot (0)$$
$$= (-2) \cdot (3 + -3)$$
$$= (-2) \cdot (3) + (-2) \cdot (-3)$$
$$= (-6) + [(-2) \cdot (-3)]$$

Therefore, $(-2) \cdot (-3) = 6$.

Fig. 7.5. Product of two negatives by use of distributive property

In the above four figures, we assume that we know how positive numbers and 0 behave under addition and that $x \cdot 0 = 0$ for all x, as proven for equation (5) in chapter 6.

Several details need to be doctored up a bit in both the pattern argument and the axiomatic argument using the distributive property. But the pattern case and the axiomatic one using the distributive property have more in common. Of course we have argued for what happens in specific cases rather than in a general proof, but even if we had generalized our presentation, a more troublesome matter applies to both approaches: wishful thinking. That is, we have no reason to argue that the pattern in figures 7.1 and 7.2 must continue as suggested. The pattern of the question marks there could have reversed their direction and mirrored what took place before the multiplication by 0 in both cases.

Similarly, we have no reason to believe that as we extend our system to include multiplication by negatives, the distributive property must operate as it did for the nonnegative numbers. We forced the distributive property (and others) to apply in our new setup in order to reach "accepted" results for multiplication of all integers.

After using an argument that depends on extending a pattern in a particular way, and after using a "proof" that assumes the validity of the distributive property in the set of integers, we are left with an interesting question about the logical connection of the two approaches. That is, does assuming that one of the two approaches is an acceptable extension in the set of negative integers affect the validity of the other approach? That is, if we assume that the pattern holds, does that legitimize the extension of the distributive property to cover the integers as well? And vice versa. Is the legitimacy of the pattern approach inherited if we accept the distributive property extension?

For a start, meeting with a few other people to figure out what exactly is being asked might be worthwhile. That is frequently half the battle in coming up with an answer. Keep everything at the level of specific numbers, which you can generalize later if you wish. For example, assume that the three bottom question marks in figure 7.2 are 2, 4, 6. Then how would you show that the distributive property in figure 7.5 holds? (Notice where exactly the distributive property is being forced to hold in that figure.) Then if you assume the argument in figure 7.5, how would that enable you to conclude that the question marks in figure 7.2 must be filled in as we did for the extension of pattern argument?

I explore other issues related to extension of negative integers in Brown (1969).

We have focused here on multiplying positives and negatives within the set of integers. How would such "proofs" hold up if we were interested in including the set of fractions? Here we compared an axiomatic and a pattern argument, not focusing on extendibility to "larger" sets. See Wu (2011) for questions regarding the limitations of extending such "rules" of multiplication to the set of fractions.

But this section intended to show that an intuitive pattern argument (as in figs. 7.1 and 7.2) and an axiomatic *perspective* (figs. 7.4 and 7.5) are essentially the same, even though they appear to be on different levels of abstraction.

On Primes

First let's review well-known and more esoteric properties of a subset of natural numbers: primes. We will then examine an even smaller subset to find greater appreciation for at least one property of primes that has been taken for granted. We begin with a recounting of a famous proof of the infinity of primes.

A Brief Verbal Proof

What follows is a slightly modernized version of Proposition 20 in Book IX of *Euclid's Elements*. Briefly, and without much symbolism, Euclid assumes that a finite number of primes exists. He then creates a new number from those primes and uses it to prove the existence of some prime that has not already been accounted for:

- He creates a new number from all the existing primes. It is the product of all those primes with the number 1 added.

- That new number is either prime or composite.

- If it is prime, it is larger than any previous prime (with the interesting assumption that multiples of natural numbers greater than 1 are greater than any natural number).

- If composite, it must be divisible by a prime number (by definition).

- But none of the primes he used to create the new number will work (because they all leave a remainder of 1 when they divide the new number he created).

- Therefore, a new prime number—other than the ones that have composed the supposed finite set of primes—must exist.

A Symbolic Rendition

Euclid's proof assumes that the number of primes is finite and shows that this assumption leads to a contradiction. The first few primes are 2, 3, 5, 7, 11, and 13. Let us denote the first one by p_1 (so $p_1 = 2$), the second by p_2 ($p_2 = 3$), and so on. Let n primes be denoted by

$$p_1, p_2, p_3, p_4, p_5, \ldots, p_n,$$

where p_n is the last prime.

Euclid creates a new number, K_n, by multiplying the finite list of primes and then adding 1:

$$K_n = (p_1 \cdot p_2 \cdot p_3 \cdot \ldots \cdot p_{n-1} \cdot p_n) + 1$$

This number is certainly bigger than any of the enumerated prime numbers. (Why?) Look closely at K_n. K_n either is or is not prime. Let's review both possibilities.

If K_n is prime, we have formed a new prime (K_n itself) larger than any of the previously enumerated primes $p_1, p_2, p_3, p_4, p_5, \ldots, p_n$, and we therefore cannot claim that these exhaust all the possible primes.

If K_n is composite (i.e., not prime), then it must be a product of primes and hence divisible by some prime. But K_n cannot be divisible by p_i for all the primes selected. Why?

$$\frac{(p_1 \cdot p_2 \cdot p_3 \cdot \ldots \cdot p_{n-1} \cdot p_n) + 1}{p_1} = (p_2 \cdot p_3 \cdot \ldots \cdot p_{n-1} \cdot p_n) + \frac{1}{p_1}$$

And we thus have a remainder of 1. For the same reason, K_n cannot be divisible by p_2. The same claim can be made for $p_3, p_4, p_5, \ldots, p_n$. But some prime other than those we have listed must divide K_n.

Brief Commentary on the Proof

This proof is one of the briefest in *Euclid's Elements*. It is *clever* in that he creates a number (K_n) that appears to come from out of the blue, and at first blush the status of that number is unclear. The proof is quite elegant and slippery, however. To follow the logic may take several readings, and even when it is carefully followed, it needs more careful analysis.

We will consider some of these issues. Why is it elegant? Look at the new number Euclid creates to show that the number of primes is infinite. If you think of the new number as being created in some temporal order, before 1 is added, it involves creating a number that could not be more unprime—being divisible by all the other primes we started with.

What does adding 1 do to that number? First, let's figure out the logic of that new number. What does Euclid claim about the new number? It is tempting to read this proof quickly and to conclude that the new number created *is itself prime*. The values for K_i in table 7.1 support this belief.

Table 7.1
Prime status of the K_is in Euclid's proof

i	K_i	=	Prime?
1	$2 + 1$	3	Yes
2	$2 \cdot 3 + 1$	7	Yes
3	$2 \cdot 3 \cdot 5 + 1$	31	Yes
4	$2 \cdot 3 \cdot 5 \cdot 7 + 1$	211	Yes
5	$2 \cdot 3 \cdot 5 \cdot 7 \cdot 11 + 1$	2,311	?
6	$2 \cdot 3 \cdot 5 \cdot 7 \cdot 11 \cdot 13 + 1$	30,031	?

The values of K_i escalate quickly. We have not discussed efficient procedures for determining whether a number is prime, but you can probably figure out why K_4 is prime and the same for K_5. To get a feel for relatively straightforward procedures to determine whether a number is prime, get together with someone and see what you can determine for K_6. It will be worth the effort. In fact it is not prime, and you need not try too many divisors to find out.

The Distributive Property

When we divide $(p_1 \cdot p_2 \cdot p_3 \cdot \ldots \cdot p_{n-1} \cdot p_n) + 1$ by any p_i, we get a remainder of 1, but that argument hides the role of the distributive property. Chapters 1 and 2 discussed what it means for a number to divide another number. That is, if a divides b, a number c must exist so that $a \cdot c = b$. So if a divides b, and a divides c, then numbers d and e exist so that $a \cdot d = b$ and $a \cdot e = c$. Then by the distributive property,

$a \cdot d + a \cdot e = a \cdot (d + e) = b + c$. And from this last equation, what flows? Without offering a proof, I should note that the application of divisibility for Euclid's proof is different from what we framed above. I leave it to you to show that p_v divides ($p_1 \cdot p_2 \cdot p_3 \cdot \ldots \cdot p_{n-1} \cdot p_n$) for every prime and that p_v does not divide 1, and therefore p_v does not divide ($p_1 \cdot p_2 \cdot p_3 \cdot \ldots \cdot p_{n-1} \cdot p_n$) + 1 (i.e., x divides y and x does not divide 1; therefore, x does not divide $y + 1$).

Other Prime Matters

Most people have had experience with matters that flow from the concept of prime, even if they have not seen proofs. For example, in grade school, people learn to factor numbers into a product of primes. How would you break 210 up into a product of primes? Suppose you start with $42 \cdot 5$, and I start with $7 \cdot 30$. If we each continue factoring, what do we get?

> You: $42 \cdot 5 = 21 \cdot 2 \cdot 5 = 7 \cdot 3 \cdot 2 \cdot 5$
>
> Me: $7 \cdot 30 = 7 \cdot 2 \cdot 15 = 7 \cdot 2 \cdot 5 \cdot 3$

Perhaps a few more trials would help, but what becomes clear from many trials is that aside from order of factors, we will both end up with the same factorization: $2 \cdot 3 \cdot 5 \cdot 7$. This is known as the unique factorization property. It is considered so obvious that people who first learn about it often wonder what the big fuss is about.

Another, related matter is that of reducing fractions to "lowest terms" by factoring the numerator and denominator into a product of primes. So I might begin by reducing $12/36$ to $6/18$, and you might choose to reduce it first to $2/6$. Nevertheless, each reduction can be enticed one more step and both of us eventually get the same number, $1/3$, as the reduction of $12/36$ to lowest terms. That means not only that we both end up with the same fraction eventually but also that, just as significantly, neither of us goes on forever trying to reduce these fractions further (hence "reducing to lowest terms").

Other observations and questions we asked about primes in chapter 2 (under "'Prime' Concern") were probably less familiar than the factoring discussed here. We have already answered one question raised there: how many primes are there? But I have not said much about the others, which included the following: Can a formula generate these primes? All of them? An infinite number but not necessarily all? Do the primes distribute themselves in some interesting manner? Can we determine, without laborious calculation, whether a number is prime? We will explore some of them here. See Brown (1987) for others.

We also observed that some primes came as pairs of numbers, as in figure 7.6.

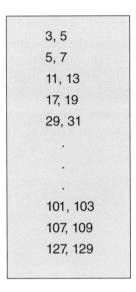

Fig. 7.6. Some twin primes

These are called twin primes. How many twin primes do you think exist? Is the number of twin primes finite or infinite?

Also, look at these equalities in figure 7.7, depicting examples of Goldbach's conjecture.

$$4 = 2 + 2$$
$$6 = 3 + 3$$
$$8 = 5 + 3$$
$$10 = 5 + 5$$
$$12 = 5 + 7$$
$$14 = 7 + 7$$

Fig. 7.7. Goldbach's conjecture for prime numbers

Figure 7.7 shows that the first few even numbers can all be expressed as the sum of two prime numbers (sometimes in more than one way, such as other ways of expressing 10 and 14 than fig. 7.7 shows). Can every even number be so expressed? German-born Russian mathematician Christian Goldbach (1690–1764) formulated that question in 1742.

On Math and Humor

Some of these questions and observations have humorous answers. We will now discuss two of them. We will return to some of the others in "Primes: Getting Even," where we will rejuvenate the role of the distributive property. The search for a formula that would generate only prime numbers enticed many well-known mathematicians in the mid-seventeenth and early eighteenth century, including Fermat, Mersenne, and Euler. Fermat conjectured that all numbers of the form $2^{2^n} + 1$ generated a prime. The formula works for $n = 1, 2, 3$, and 4 but breaks down for 5, yielding 4,294,967,297.

In the first half of the twentieth century, some progress occurred in seeking a formula to generate primes and only primes. N. H. Mills, in 1947, came up with the following formula, which he proved will generate a prime for every value of n:

$$\left[A^{3^n}\right]$$

That is, he proved that some fixed number A exists such that each time n is plugged into the formula, we will get a prime number. The meaning of the square brackets is simple: $[x]$ stands for the greatest integer less than or equal to x, so $[2.57] = 2$; $[4.2] = 4$; $[1.9999] = 1$. We had a solution at last to the search for a formula that would generate primes. Unfortunately, to make proper use of the formula, we have to know the value of A. What is it? Answer: though much effort went into producing the above formula, it reveals nothing about the nature of A. A might be some real number (e.g., $\sqrt{2}$) between 1 and 5, or it could be larger than the number of stars in the universe.

Why is that funny? The incongruity between what we might expect for an answer to some practical question in our daily lives versus what we seem to be presented with here is enormous. Say that you wanted to know when expected guests would arrive for dinner. Someone tells you that the guests would undoubtedly arrive at some definite time, but there was no way to determine what that time was. Because it might be within the hour or within the decade, you would think that information ludicrous.

Sometimes mathematical exploration simultaneously engenders both mathematical beauty and frustration. The culprit often is the assertion that "there exists" something, even though we may have no idea of its location or how to create it. It is a kind of statement often used with pride in mathematics but is used sparingly in most other contexts. We do know of such real-world "existential" statements, but they usually are not helpful. For example, we know (as an empirical, inductively arrived-at fact) that no matter how healthy we may be, at some precise moment we will die. That realization may direct how we live our lives, though it is something we tend to repress. Without more specifics, it is difficult for such information to have much effect on how we live in any fine-grained way.

Why does (can) mathematics arrive at such existential statements—especially ones that do not specify the strategies for creating the object they specify? The answer is that the special form of proof known as proof by contradiction depends on showing that denying the existence of certain elements or conditions would produce a contradiction within the system.

A particularly humorous example of a mathematical existential theorem derives from an effort to prove the above-mentioned famous conjecture by Goldbach depicted in figure 7.7. Goldbach probably came upon his conjecture by doing something quite clever. He may well have decided to stand a humdrum observation on its head—something that others may have observed but ignored as insignificant. I would guess that he began with the observation that because all primes greater than 2 must be odd, you will always get an even number if you add two primes (excluding 2). That is surely an unremarkable observation.

What turned out to be a brilliant move, however, may have depended on Goldbach's disinclination to discard the above observation about the sum of any two primes greater than 2. He may well have transformed a dull observation into a mind-blowing one by performing a "what if not" on the simple observation, which led to something that puzzled mathematicians for a long time to come.

Though we can come up with many instances to substantiate his conjecture (and you may be able to continue the pattern for quite a while), no one made any headway with a general proof for almost two hundred years. Then in 1931, another Russian mathematician, Lev Schnirelmann (1905–1938), produced the first crack in proving a more modest form of the conjecture. To appreciate what he proved, recall that Goldbach wanted to establish that for any even number, two primes would add up to that given number. His conjecture focused not on finding or showing that many different pairs could work but rather on that at least one pair existed for any even number.

What did Schnirelmann do? Because Goldbach was hoping to show that at least one pair could always be found, what would be a really funny statement and proof regarding how many primes you might need to add to achieve a given even number? Perhaps ten would be funny, one hundred, or one thousand. Would the need to add three hundred thousand natural numbers instead of a pair create a belly laugh? That was Schnirelmann's contribution. He showed that given any even number, we can find at most three hundred thousand primes that must be added to achieve it.

Schnirelmann did not achieve his goal by acquiring evidence inductively. It was not an act of trial and error but a logical deductive proof, the same concept of proof that Goldbach sought (even though the focus on three hundred thousand as opposed to two was a bit much).

The enormous incongruity between what is sought as a solution and what is produced is sometimes funny. Imagine that Barack Obama wanted to gather a close-knit clan of colleagues to develop a workable plan to find and capture Osama bin Laden. Further imagine that Obama's most trusted and wise consultant said he probably would need no more than twenty-five people. Then Obama set up a team to spend a year determining the names and positions needed for this mission to have a high probability of success. Suppose the team then came back and told him that after much exploration, they discovered that they would need three thousand people instead of twenty-five. Here of course the analogy with Goldbach and Schnirelmann breaks down a bit because the manner of proof was not one of logical deduction, but involved sophisticated strategies to gather empirical evidence (with perhaps a bow to theoretical strategies of probability).

Nevertheless, Schnirelmann's discovery was valuable for giving some headway on a problem that had been lying fallow for a long time. It was a first approximation, and the humor of it need not detract from the fact that it, and many problems like it, was successfully solved by making small iterations in both their solution and in the way the problems are posed.

Primes: Getting Even

Many of the questions we asked about prime numbers, some of which have been unanswered for centuries, take on a new complexion if we modify the domain from the natural numbers to some other set. The extension uses the same definition of prime number but applies it to a different domain. Investigation began in the set of natural numbers: N = {1, 2, 3, 4, 5, 6, 7, . . .}. Look instead at the set E = {1, 2, 4, 6, 8, 10, . . .}. E denotes the set of even numbers (with 1 thrown in, as in the set N). A key element in transforming the nature of answers to many of the questions in E will often depend on a dominant role of the distributive property, though it is not an exclusive component.

To start, we may ask what some of the primes are in set E. A first, quick answer might be that the only prime in E is 2 because all the others are even numbers and therefore are divisible by more than two numbers. For example, one might claim that 6 is not prime because it is divisible not only by 1 and 6 but also by 2. By analogy, that reasoning would suggest that 5 is not prime in N because 2 divides 5. Though 2 is in N, we reject that conclusion because 2.5 does not belong to N. This correction clarifies how divisibility depends on the set within which it takes place. The claim that 2 divides 6 in E is based on the observation that $2 \cdot 3 = 6$. But 3 does not belong to E in the same way that 2.5 does not belong to N. To say that a divides b if there is a c so that $a \cdot c = b$ assumes that all the elements belong to the same set.

With this clarification, the primes in set E are 2, 6, 10, 14, ... Here 6 is prime because only 1 and 6 divide it. Why do none of these numbers have divisors other than 1 and themselves? The composite numbers in E are 4, 8, 12, 16, ... Each element has more than two divisors in E. Why?

So after reflection (and we can come to see this in many ways), several prime questions that we investigated in N—and that were problematic—essentially collapse onto one (or almost one) lens. Primes are numbers that can be expressed as 2 × an odd number. We can designate such numbers as $2 \cdot (2 \cdot n - 1)$. That expression in one fell swoop lets us determine (1) how many primes are in E, (2) whether a number in E is prime, and (3) a generating formula for all primes in E.

Getting a formula for all E's primes makes us that much more envious of something we never accomplished in doing so for set N. Here we have a formula that not only generates all the primes but also does so in order, capturing them all in succession. The formula we found for generating and characterizing primes in E could be expressed slightly differently from $2 \cdot (2 \cdot n - 1)$. Using the distributive property:

$$2 \cdot (2 \cdot n - 1) = 4 \cdot n - 2 \text{ for all } n \geq 1 \text{ in } N$$

Now we move on to unique factorization and reducing fractions to lowest terms.

(1) Notice what happens when the following numbers in E are reduced to a product of primes in that set:

(a) $36 = 2 \cdot 18 = 6 \cdot 6$ (not unique)

(b) $60 = 2 \cdot 30 = 6 \cdot 10$ (not unique)

(c) $20 = 2 \cdot 10$ (unique)

(d) $12 = 2 \cdot 6$ (unique)

(2) Look at the following fractions reduced to lowest terms (using elements of E as factors):

(a) $\dfrac{12}{20} = \dfrac{6 \cdot \cancel{2}}{10 \cdot \cancel{2}} = \dfrac{6}{10}$ (only one way to reduce to lowest terms)

(b) $\dfrac{12}{36} = \dfrac{6 \cdot \cancel{2}}{18 \cdot \cancel{2}} = \dfrac{6}{18}$ (a first way to reduce to lowest terms)

(c) $\dfrac{12}{36} = \dfrac{2 \cdot \cancel{6}}{6 \cdot \cancel{6}} = \dfrac{2}{6}$ (second way to reduce to lowest terms)

It appears that in (1) and (2), unlike the story of factoring numbers from N and reducing fractions with numerators and denominators from N, we do not always end up with unique factorization or unique reduction of fractions in E. Using the definition of prime in E, can you figure out the cases for which unique factorization breaks down? Similarly for reduction of fractions to lowest terms, can you figure out when a fraction will yield lowest terms in more than one way? To be clear about the meaning of "lowest terms," notice that for (2) (b) and (c), $\frac{6}{18}$ cannot be further reduced, for E has no common denominator that the numerator and denominator share; the same is true for $\frac{2}{6}$, even though $\frac{2}{6}$ and $\frac{6}{18}$ are equal.

Now with regard to Goldbach's conjecture in E:

(3) Observe the numbers to the left of the equals signs in (a)–(d); then look at the pairs of numbers on the right-hand side.

   (a) $6 = 2 + 4$

   (b) $8 = 2 + 6$

   (c) $10 = 2 + 8$

   (d) $12 = 6 + 6 = 2 + 10$

Which of the above represents instances of Goldbach's conjecture in E? For a start, we are seeking even numbers on the left-hand side and pairs of primes on the right. We are in trouble on the left-hand side in two cases. What are they? Recall that we are focusing to begin with on even numbers on the left-hand side. But (a) and (c) are automatically excluded because the numbers on the left side in (a) and (c) are not even in E. To be even, the number must be divisible by 2, but clearly 2 will not divide 6 and 10. Why not? What are the first few even numbers in E? The first is 4; the next is . . . ?

Let's look at (b) and (d) now. The left-hand-side numbers (8 and 12) are indeed even. Now look for pairs of primes of E in (b) and (d): 2 and 6 is a legitimate pair for (b); for (d), we have two pairs of primes and each works.

How would you express even numbers in E? Using $2 \cdot n$ for n belonging to N would not work. Why? But $4 \cdot n$ for n belonging to N would work. Starting with any such even number, you could get a prime in many ways: either $4 \cdot n - 2$ or $4 \cdot n + 2$ would work.

If you were to hypothesize that Goldbach's thinking in E might have been inspired as we imagined it might have been in N, then you would have started by exploring (and perhaps quickly dismissing) the converse of his actual conjecture: that the sum of two primes in E is an even number. If you had two primes in E, $4 \cdot k - 2$ and $4 \cdot j - 2$, what would you get when you add them? A modest application of the distributive property would yield $4 \cdot (k + j - 1)$. So we would have an even number when adding them.

Now suppose you become Goldbach and create the converse. Given any even number, can you represent it as the sum of two primes? If you are looking for only one pair that will work for each even number, do you see something that might emerge—starting with results that would work for 8 and 12 [(b) and (d) above]? What would work for 16? 20? If you start with an even number, $4 \cdot n$, what are two primes that work? If you start with 2 as the first, how would you express the other?

Interestingly, 12 [from (d)] will work not only for 2 plus some other prime (which?) but also for $6 + 6$. What would the next even number be that would yield two other primes?

You now might want to investigate the twin prime conjecture in E. Remember that in N, the conjecture was that an infinite number of twin primes exists. If 3, 5; 5, 7; 11, 13; 15, 17 are the first few twin primes in N, what would be the first few twin primes in E? Can you prove that the set would contain an infinite number of twin primes (something that has been such a stickler in N)?

Geometric Expansions

Chapters 1 and 2 explored the geometric version of the distributive property. We indicated that Euclid could state (in a modified way, as we will soon see) and prove the famous theorems of Euclidean geometry, even though he had no way to measure areas and lengths with associated numbers.

The eventual evolution of Euclid's worldview to a modern one that uses real numbers extensively (with algebraic formulas and numbers for areas of geometric regions) is a fascinating story—one that awaited an understanding of real numbers (such as the length of the diagonal of a square of unit length) to make it all legitimate. That is, it was necessary to see numbers such as $\sqrt{2}$ as viable measures for areas and lengths of lines.

As much as the modern view required putting irrational numbers on a firm setting, that transition neglects to portray the brilliance of Euclid's view—one that is too cavalier in dismissing the brilliance of Euclid's reliance on purely visual geometric transformations in the absence of an understanding of the set of real numbers.

I previously claimed that Euclid's worldview was in some sense childlike. But he made some clever leaps—ones that a neophyte might not readily produce. With enough patience, someone with a child's view of areas and an appreciation for cutting and pasting could follow much of what he had done.

Here we will explore these issues by using perhaps the best-known mathematical treasure: the Pythagorean theorem. When I ask grown-ups who studied geometry in their youth what they remember most vividly, they invariably recall that theorem. If they recall much about the theorem, it is usually an algebraic rendition, and furthermore they see it in a narrow venue.

We will alternate between a modern view and Euclid's conception of more than two millennia ago—to indicate how easy it is for modern readers to unflinchingly adopt a numerical mentality even when trying to demonstrate a purely geometric one.

This section is both chock full and slow moving. Do skip through details if you find them distracting, and reexamine them after the big picture becomes clearer. The pace is slow because I am trying to invite you to see the unveiling of the proof in a way that relives what I imagine could have been Euclid's approach. I am essentially invoking a concept that I speak of in chapter 8 as "pseudohistory."

The Pythagorean Theorem: A Modern Version

In modern language, what does the Pythagorean theorem say? One version asserts that in a right triangle with sides of length a and b and hypotenuse of length c, $c^2 = a^2 + b^2$. Hundreds of different ways of proving it exist. See Loomis (1968). Figure 7.8 shows one.

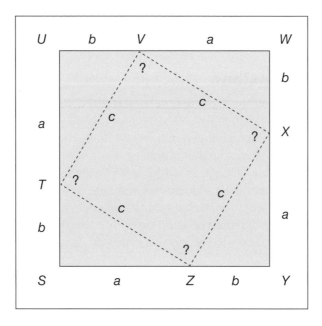

Fig. 7.8. An algebraic scheme to prove the Pythagorean theorem

Consider any right triangle, TUV, with sides of length a and b and hypotenuse of length c. Make a square, $SUWY$, of length $a + b$, so that three more congruent triangles are constructed around TUV, per figure 7.8. All angles designated by "?" are right angles because each is surrounded by two angles that add up to ninety degrees. Why? Each triangle has one right angle and two others that add up to a right angle.

The area of the big square is now equal to the area of the four triangles plus the little square within. Adding up the areas of the four triangles and the square in the middle, we get

$$(a + b)^2 = c^2 + [4 \cdot (\tfrac{1}{2} \cdot a \cdot b)] = c^2 + 2 \cdot a \cdot b.$$

But, returning once more to the famous distributive property, we can also express $(a + b)^2$ as $a^2 + 2 \cdot a \cdot b + b^2$. Therefore, $c^2 + 2 \cdot a \cdot b = a^2 + 2 \cdot a \cdot b + b^2$. Subtracting $2 \cdot a \cdot b$ from both sides, we have $c^2 = a^2 + b^2$.

Euclid's Formulation

Euclid offered two proofs of the Pythagorean theorem. Let's look now at the first: Book I, Proposition 47, in which he states, "In right-angled triangles the square on the side subtending the right angle is equal to the squares on the sides containing the right angle" (Heath 1956, p. 349).

The shading of figure 7.9 suggests how the proof will proceed. As the dotted line indicates, the proof begins by dropping a perpendicular line from the right angle of the triangle to the hypotenuse, a, and this line extends to the base of the square on the hypotenuse. It breaks the hypotenuse into two segments, x and $a - x$. He then shows that the two lighter gray figures (rectangle and square) have the same area and that the two darker gray figures (also rectangle and square) have the same area. (I am encouraging some flexibility of thought by choosing the hypotenuse to be a and the legs to be b and c.)

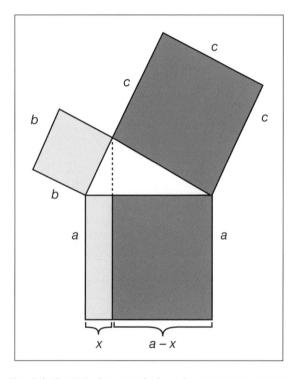

Fig. 7.9: Euclid's first proof of the Pythagorean theorem

How do we proceed? If the same shadings have the same areas, then—

(i) $a \cdot x = b^2$ for the lighter gray figures;

(ii) $a \cdot (a - x) = c^2$ for the darker gray areas; and

(iii) the distributive principle in (ii) then leads to $a^2 - a \cdot x = c^2$.

So, substituting b^2 for $a \cdot x$ in (iii), we arrive at $a^2 = b^2 + c^2$, the Pythagorean theorem in figure 7.9.

A Fast One: A Retreat

You may have noticed that we just pulled a "fast one"—actually, two fast ones. First, we based our proof on the fact that the two same-shaded regions had equal area. Second, we based our calculation on the lengths of the various segments, *a*, *b*, *c*, *x*, and *a − x*. The proof used numbers for the lengths (and numbers for the areas)—something Euclid did not have at his disposal.

In the spirit of calisthenics, let's look at several examples to demonstrate the mindset needed to deal with areas according to Euclid. The first encounter: suppose you have a parallelogram (fig. 7.10), a four-sided figure with opposite sides parallel.

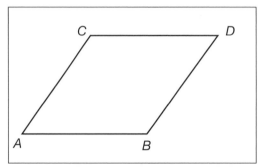

Fig. 7.10. Parallelogram with area to be explored without use of numbers

Without resorting to numbers, can you create a rectangle that has sides *AB* and *CD* in common with the parallelogram but that has the same area as the parallelogram? Reminded of the connections between the distributive property and its geometric form, you probably can create a rectangle that has the same area as the parallelogram in less than a Superman leap. Allowing for the sloppiness (as we did) of not explicitly stating all that is conveyed in the diagram, can you show that the rectangle has the same area as the original parallelogram? Consider figure 7.11.

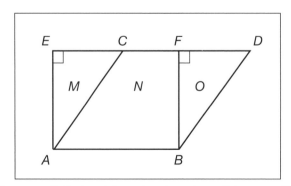

Fig. 7.11: "Same area" displayed for rectangle and parallelogram without numbers

But how have we proven that the parallelogram and rectangle have the same area without using area formulas?

As indicated in figure 7.11, both the rectangle and parallelogram were composed of two congruent triangles (*M* and *O*) with a common region, *N*. So, we can show that two figures have the same area by indicating that they are composed of congruent regions and are joined to a common region—something we can achieve with cutting and pasting. I am not alluding to theorems needed to prove why the triangles are congruent, but careful cutting and pasting would at least convey what is intuitively obvious. We are overlooking here a proof that the two triangles are congruent—but taking that for granted, we can see how the metaphors of cut and paste led us to that conclusion.

How could you make the above description more precise? What assumptions have we made about joining congruent regions? I leave those activities for you, but the pictures convey that this is an activity that youngsters could follow.

Essentially, we are putting together regions and have assumed that (1) two congruent regions have the same area and (2) two regions that can be reduced to congruent regions (placing them together along line segments, as in the above example) also have the same area.

Try a second example that adopts the same mindset. This one is a bit more complicated but uses similar strategies. In figure 7.12, how do the areas of △*ABC* and △*ABD* compare?

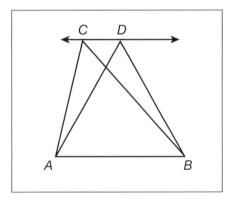

Fig. 7.12. Two triangles whose areas are to be shown equal without using formulas

If △*ABC* and △*ABD* are on the same base and each has its third vertex (*C* and *D*, respectively) on a line parallel to the base, can you show that they have the same area? How would you do it by using a formula for the area of a triangle?

Now try to show that the two triangles have the same area without associating numbers for their sides or altitudes. This approach may be a bit trickier than what figure 7.10 called for. One way might be to look at each triangle separately. We already compared the areas of a parallelogram and an associated rectangle in figures 7.10 and 7.11, even if we cannot come up with a numerical evaluation. To start, look at △ABC, and see whether you can find a parallelogram that has the same base as △ABC with its second base on a line parallel to it.

Once you do that, look at figure 7.13. How does the area of △ABC compare with that of the associated parallelogram ($ABEC$)? Because △ABC is congruent to △BCE, △ABC is one-half the area of its associated parallelogram $ABEC$.

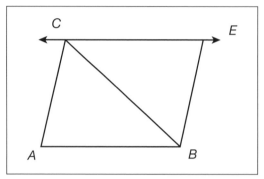

Fig. 7.13. Intermediate step to show that the two triangles in figure 7.12 have the same area

Look once more at figure 7.12. Having found an associated parallelogram for △ABC, how might you create one for △ABD, using the same concept of "associated parallelogram"?

I leave it to you to sketch the two associated parallelograms for the two triangles in figure 7.12. Essentially duplicating the reasoning we used for figure 7.11, we can see that the two associated parallelograms have the same area. Because each of those areas is twice that of the original triangles, we see that the triangles have the same area.

Big Picture So Far

So far we have seen that △ABC and △ABD in figure 7.12 are equal in area even though (1) the two triangles are not congruent, (2) we are not using a formula for the area of a triangle, and (3) we do not have a number associated with that equal area.

It may look as if we violated the condition that we must be able to show that figures have the same area if we can cut and paste them together so that each is composed of identical or congruent pieces. Have we done that here? Look back at both triangles in figure 7.12. In each case, I have tried to convince you that they are equal in area to half their associated parallelogram and that both associated parallelograms are equal in area. Does claiming that each of them is half the area of its associated parallelogram mean that we have, alas, gone back to requiring numerical information—something I claimed that Euclid avoided? Notice, however, that in claiming that each triangle is equal to half of the parallelogram, we have implicitly done a bit more. That is, the two triangles that compose each of the areas of their respective parallelograms are in fact congruent. I leave it to you to untangle that apparent hitch. Check with a friend or colleague if this is a problem (which of course is a good thing) not easily overcome.

Back to Euclid and the Pythagorean Theorem

Though we have engaged in calisthenics that bear directly on Euclid's first proof of the Pythagorean theorem, we will not go through all the details. But I will allude to what needs to be done to put the pieces together. It is not complicated conceptually, but the diagrams are a bit hairy looking until you play with them for a while, and they are best understood if you try it yourself.

We acted shiftily when we came up with an algebraic solution in figure 7.9 that depended on the assumption that the two light-gray regions were equal in area and the same was true for the two dark-gray regions. But we finessed a proof that this was the case, taking that for granted by how the figures were separated.

Let's look back at a less classy-looking sketch of figure 7.9. Figure 7.14 is the original sketch of the three squares on the sides of a right triangle.

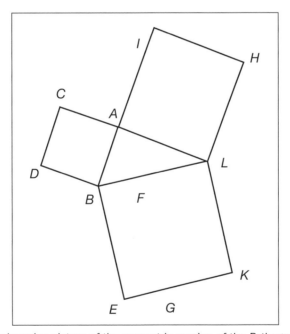

Fig. 7.14. Noninvasive picture of the geometric version of the Pythagorean theorem

Let's look now at the scheme that Euclid used in Book I, Proposition 47, of *Euclid's Elements*. We will be adding a bit more than shown in figure 7.9. You may want to look back to see what has been added. It is the crux of his proof as sketched in figure 7.15. The three extra lines added to figure 7.14 enable him to show that the two light-gray regions in figure 7.9 are equal in area.

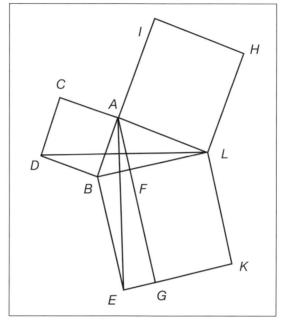

Fig. 7.15. Crux of Euclid's proof with three extra lines drawn

We have the perpendicular line from the right angle (angle A) of $\triangle ABL$ through both sides of the square on the hypotenuse, as we had in figure 7.9. But Euclid constructed two other lines: AE and DL. Here's where things may look confusing, but they should soon clear up. Euclid focused on two triangles now formed by his two new lines. He chose to look at $\triangle ABE$ and $\triangle DBL$. If you look at both triangles, you may notice many things. Why would you guess he chose them in particular? Two sides of one triangle are congruent to two sides of the other. With a little squinting, you can see that the two triangles singled out also have two equal angles (look for obtuse ones). These two triangles are then congruent (having two sides and an included angle congruent).

So what? Look back at figure 7.9; what did we take for granted when we came up with an algebraic proof? It was essentially that the two light-gray rectangles and the two dark-gray rectangles have equal areas. Here, to eliminate clutter, I focused on only the equal area of the two light-gray regions in figure 7.9 and have not tackled the dark-gray regions of that figure.

If we could create two additional similarly situated triangles in figure 7.15 (lines BH and AK) associated with the regions of figure 7.9, we could capture the equality of the dark-gray regions in figure 7.9 too. Voilà, the Pythagorean theorem would be demonstrated—almost. We left out an assumption: that something is significant about the two congruent triangles $\triangle ABE$ and $\triangle DBL$. For this proof to work, what might that significance be? Look again at figure 7.9 and try to guess which triangle is associated with the light-gray square and which is associated with the light-gray rectangle.

Before reading on, try a reasonably accurate and large sketch of figure 7.15. Then act in a way that depicts what I have described as Euclid's inclination—like a child. Cut and paste $\triangle DBL$ and $\triangle ABE$.

Considering that I have left out the sketch of the analogous triangles associated with the two dark-gray regions, try to figure out which triangle in figure 7.15 is associated with figure 7.9's rectangular region ($BFGE$) and which is associated with the square ($BACD$) of figure 7.9. If this question sounds confusing, recall how we spoke of parallelograms associated with triangles in figure 7.12 and 7.13. Now compare rectangle $BFGE$ with $\triangle ABE$. Part of the confusion here is that $\triangle ABE$ is not included completely within that rectangle. Another difficulty is with the altitude to the side of the triangle. The next section will clarify this thought.

Some degree of playing around, as we did earlier for the calisthenics, will indicate a powerful connection between each triangle and its associated rectangle. The cutting and pasting of the two triangles to show that the area in each case is significantly related to areas of square $BACD$ and rectangle $BFGE$ may be easier by first using the (forbidden for our purposes here) area of a triangle in relation to the length of a side and its associated altitude. At any rate, if you want to anticipate the grand finale, jump to the calisthenic exercise of figure 7.16 (discussed in the next section), which claims that the area of $\triangle ABE$ is half that of its associated rectangle $BFGE$.

Apologies and Almost Finale

That last section did not give the complete visual resolution (showing why each significantly created triangle is exactly half the area of the associated rectangle) because I wanted you to play around as a child might. I will give you a hand in the finale, however, by isolating one of the triangles in figure 7.15 ($\triangle ABE$) and showing which rectangular region it relates to as a final calisthenic exercise. See figure 7.16. Can you do some cutting and pasting (and perhaps extend the lines in the diagram analogously to what we did to show that both triangles in fig. 7.12 are congruent)? I indicated that its area is half the area of the corresponding rectangle. Try extending the altitude from A to BE and see whether you can show that the area of $\triangle ABE$ is half that of rectangle $BFGE$.

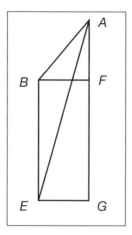

Fig. 7.16. The area of △*ABE* is half that of rectangle *BFGE*.

It would probably be relatively easy for you to follow the cut-and-paste argument if it were handed to you. It took a genius to (1) split the hypotenuse into two rectangular sections, (2) show that each split region on the hypotenuse had the same area as the squares on the sides, and (3) use triangles embedded within the regions of the rectangles as "middle people" to demonstrate what the light-gray and dark-gray regions in figure 7.9 suggest.

Following Euclid's proof of the Pythagorean theorem is probably like what people experienced encountering the wheel as a solution to hauling heavy objects over long distances. They probably could appreciate the invention as an act of genius even though they may not have come up with it on their own.

If you are tempted to follow the details of this proof, review it with a friend or colleague—or go directly to the original source, Book I, Proposition 47, of Euclid's three-volume set (Heath 1956, pp. 348–49). That will be much easier to read if you have begun to follow even some of what we have covered. Also, the following website does a good job of discussing the theorem and its history: http://aleph0.clarku .edu/~djoyce/java/elements/bookI/propI47.html.

Euclid's More Stupendous Rendition

We now move from Euclid's famous Pythagorean theorem exposition to his lesser-known version. It is more impressive than what we just discussed—not only for its proof but also for its very statement. As in the first proof, he does not use algebra; moreover, he does not talk about squares on sides of a right triangle, but rather something more general. In its most general sense, the Pythagorean theorem is not primarily about squares: "In a right-angled triangle, the figure on the side subtending the right angle is equal to the similar and similarly described figures on the sides containing the right angle" (Book VI, Proposition 31; Heath 1956, pp. 268–69).

Why is this version of the Pythagorean theorem a bit unbelievable? More amazing than Euclid's realizing that the theorem works for similarity is that he could cope with that concept (similarity) at all. Roughly, two figures are similar if one is a blown-up version of the other (without distortion). In modern terminology, two rectilinear figures are similar if all their corresponding angles are equal and their corresponding sides are in proportion. Thus, a modern analysis of similarity involves using the lengths of sides and ratio of geometric figures. Recall, however, that Euclid could measure neither line segments nor areas. He dealt with ratio in the same way that he dealt with area. That is, he could speak of "same area," "same lengths," and "same ratio" but could not measure them with a number.

To quickly appreciate the astounding nature of this formulation, we will jump to the twenty-first century and associate numbers with the theorem.

To introduce the power of this version of the proof, before we dispose of squares, we will hang them onto the sides of a right triangle for a final occasion.

Look at a 5–12–13 right triangle. If you put squares on the sides, then, as in figure 7.17, they have respective areas of 25, 144, and 169. Now, cut each square into rectangles so that they have the same lengths as on the three sides of the right triangle but have heights half as long. Now look at the half squares on each side of the right triangle, as in figure 7.18.

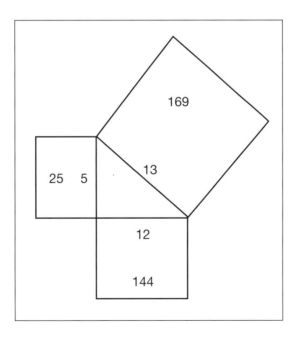

Fig. 7.17. Squares on the sides of a 5–12–13 right triangle

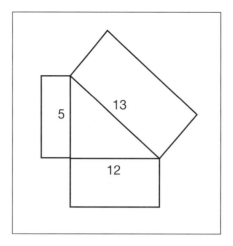

Fig. 7.18. Areas on the sides of a 5–12–13 right triangle with heights that are half the lengths

Then we have rectangles on the sides with lengths of 5, 12, and 13 and have respective widths of $\frac{5}{2}$, $\frac{12}{2}$, and $\frac{13}{2}$. What became of the Pythagorean theorem? Well, what are the areas of the three rectangles? You can calculate them individually: $\frac{5}{2} \times 5$, $\frac{12}{2} \times 12$, and $\frac{13}{2} \times 13$. Can you show easily that the areas satisfy the Pythagorean condition?

What do you get? Alternatively, you can use the distributive property and determine that if $c^2 = a^2 + b^2$, then half their areas must also be equal; that is, $\frac{1}{2}c^2 = \frac{1}{2}(a^2 + b^2) = \frac{1}{2}a^2 + \frac{1}{2}b^2$.

Was anything special about choosing the heights to be one-half of the respective lengths on the three side of the right triangle? How would you show that any fraction or multiple of the widths would work? What axiom would you use?

So it looks as if we have a modified form of the Pythagorean theorem. If you focus on the three sides of a right triangle, and then create three rectangles upon each of those lengths whose heights are the same multiple of the lengths, then the sums of the areas on two of the sides are equal to the area on the hypotenuse.

But that's only part of the story. Euclid's second proof of the Pythagorean theorem leads us even farther down the garden path. Look at figure 7.19. Around the right triangle, we have placed not only squares and rectangles but also three equilateral triangles. Still holding on to formulas at this point to gain a quick insight that we can later replace with Euclid's worldview, how would you compare the areas of the three equilateral triangles? As long as we are now deeply into use of twenty-first-century number notations, let's use the formula for the area of an equilateral triangle with side of length s:

$$s^2 \frac{\sqrt{3}}{4}$$

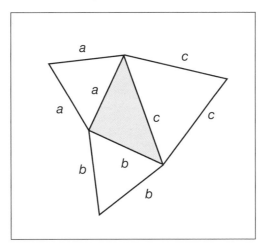

Fig. 7.19. The Pythagorean relationship demonstrated by three equilateral triangles on the sides of a right triangle

How would you now show that the three equilateral triangles operate in the same way as the Pythagorean theorem for squares (and can even be derived from the theorem for squares)? How far might you deviate from the regions in figures 7.17–19 in seeking Pythagorean generalizations based on similarity of figures? All the regions shown so far have consisted of straight lines and associated angles. Can you come up with other regions that are curvilinear (such as semicircles)? Perhaps nonsymmetrical regions that an artist might create as potential objects to generalize the Pythagorean theorem (such as three ducks of the same shape but different sizes, all sitting on a flat body of water along the sides of a right triangle)?

Place three semicircles on the sides of a right triangle. Using the formula for the area of a circle (πr^2 for r, the radius), try the Pythagorean relationship. With that as our last hurrah on the Pythagorean theorem, I will go no further with details and quasiproofs, except to point out that by taking a primitive nonnumerical geometrical view of areas, Euclid came to understand the Pythagorean theorem in terms considerably more general than the standard depiction. Remember: supposedly equivalent mathematical formulations of a property as simple as the distributive property can lead us to diverse ways of generalizing the formulation.

What is astounding about his accomplishment is that without using formulas and without using much more than the equivalent of scissors and paste, Euclid could continue to speak of "same area." What makes it particularly impressive is that similarity involves proportion. Modern terminology defines two rectilinear regions as figures whose corresponding angles are equal and whose corresponding sides are in proportion. But how we deal with proportion without associating the entire set of real numbers with the sides is a story of its own. The question itself is a powerful enough end to our journey.

Retrospective

Lest we have a forest-and-tree problem, we should pull together some ideas. The first few sections of this chapter explicitly used the distributive property. Its application in demonstrating such phenomena as the property of unique factorization, plus its enabling us to show the downfall of unique factorization in a slightly different domain, was moving.

More moving, perhaps, was the paper that you cut and pasted to replicate Euclid's thinking in approaching lengths, areas, and similarity without having an adequate supply of real numbers to measure them. Today, his inability to use real numbers to measure his geometrical objects would be considered primitive. From another perspective, it was genius. As mentioned in chapters 1 and 2, algebraic and geometrical forms of the distributive property may have been equivalent logically, but for our ability to "see" and to generalize, they function differently. His approach was childlike in one sense but sophisticated in another.

References

Brown, Stephen I. "Signed Numbers: A 'Product' of Misconceptions." *Mathematics Teacher* 62: (March 1969): 183–95.

———. *Some Prime Comparisons*. Reston, Va.: National Council of Teachers of Mathematics, 1987.

Heath, Sir Thomas, ed. *The Thirteen Books of Euclid's Elements*. New York: Dover Publications, 1956.

Loomis, Elisha. *The Pythagorean Proposition*. Reston, Va.: National Council of Teachers of Mathematics, 1968.

Wu, Hung-His. "Phoenix Rising: Bringing the Common Core State Mathematics Standards to Life." *American Educator* 35 (Fall 2011): 3–13.

Further Reading

Brown, Stephen I. "Towards a Pedagogy of Confusion." In *Essays in Humanistic Mathematics*, edited by Alvin White, pp. 107–22. Washington, D.C.: Mathematical Association of America, 1993.

———. "The Problem of the Problem and Curriculum Fallacies." In *Constructing Mathematical Knowledge: Epistemology and Mathematical Education*, edited by Paul Ernest, pp. 175–89. London: Falmer Press, 1994.

———. *Reconstructing School Mathematics: Problems with Problems and the Real World*. New York: Peter Lang Publishing, 2001.

———. "Relatively and Philosophically E(a)rnest": Festschrift in Honor of Paul Ernest's 65th Birthday. In *The Montana Mathematics Enthusiast: Monograph Series in Mathematics Education*, edited by Bharath Sriraman, pp. 97–128. Charlotte, N.C.: Information Age Publishing, 2009.

Brown, Stephen I., and Marion I. Walter. *The Art of Problem Posing*, 3rd ed. Mahwah, N.J.: Lawrence Erlbaum Associates, 2005.

CHAPTER 8

Commentary for T. C. Pits

Dear T. C.

This last chapter features "T. C. Pits" (shorthand for "the celebrated person in the street"). The name is a less sexist takeoff from the title of a delightful book by Lillian Lieber (1942), *The Education of T. C. Mits* ("the celebrated man in the street"). It is one of a collection of books (many written in verse form) dealing in a popular way with such topics as infinity; Einstein's theory of relativity; and the relationships among science, art, and mathematics. My favorite high school mathematics teacher, Sylvia Somerman, a student of Lieber's at Evander Child's High School in New York City, gave me *T. C. Mits*. That book was part of an invaluable legacy that Sylvia bestowed on me—part of which included a collection of "cute and culture" math problems that I cherished because they gave hardly a clue about what strategies might be useful in thinking about them. I had taken her collection of cute and culture problems with me to summer camp when I was nineteen years old, where I first made significant headway on one of those problems in the men's room (I enjoyed the solitude). Not having access to paper and pencil (nor were computers available), but only a red marker, I sketched out the solution on the door of the bathroom stall. I wonder whether it remained there or whether anyone was tempted to obliterate it or claim ownership.

With this anecdote, I both summarize salient features of this book and briefly suggest several additional relevant directions.

Though this book explores several broad issues in mathematics and education, it does not offer a detailed critique of curriculum proposals writ large, nor does it suggest how to sequence topics within or between the chapters. It also does not define essential content to be covered over a span of school years. Rather, I intended this book to entice you with connections and ways of thinking about mathematics from a variety of perspectives. It milks one ostensibly small concept, the distributive property, far beyond its usual due. Part of the purpose was to experiment with a playful style and format rather than to offer a didactic treatise. If my approach inspires other people to push the bounds of alternative formats and styles in reflecting on mathematical concepts and their teaching, I will have achieved an important goal.

A Humanistic Orientation

Not only have I selected one element of mathematical content as a spotlight to reveal a wide variety of mathematical thinking, but I have also connected and embedded that element within humanistic dimensions.

This book has used literary devices, such as metaphors, for several purposes—one being to explore paths whose conclusions I had not anticipated. An entire chapter pounded away at the metaphor of "striving" by viewing some calculations not from their actual statement but rather as something heading elsewhere.

Not often associated with mathematical thinking, humor is usually viewed as an aside rather than as a force that enables us to appreciate achievements in thinking. This book has used humor to comment on questions and analyses that would otherwise appear strictly logical. In various sections on prime numbers in a variety of domains, for example, some theorems are quite funny compared with what we normally define as "progress."

This book has also used—and encouraged you to engage in—storytelling to explore ideas and feelings. The role of a dream becomes the driving force to investigate the significance of "same/different" as a fundamental mathematical category in chapter 5.

Underlying many of these strategies is the attempt to connect mathematics with personal and emotional reactions often treated as if they are unconnected with a strictly rational presentation of subject matter.

We will discuss elements common between this book and the standards, as well as explore differences. This book, however, is not a frontal critique of the standards but rather an invitation for dialogue that explores how these two orientations can enrich one another. It is written as much for T. C. Pits, who wishes to imagine mathematics and its educational practice in a different light. A brief review of excerpts from the Common Core State Standards for Mathematics (CCSSM; National Governors Association Center for Best Practices and Council of Chief State School Officers 2010) will serve as a helpful bas-relief.

Bas-Relief: Recent Standards

Motivated in part by the relatively low mathematics achievement of U.S. students among developed countries, CCSSM refines and redefines (1) the mathematical content and (2) student practice in grades K–12. Unlike previous agendas, the Common Core standards have been influenced significantly by individual states and are intended to foster a sense of national unity. Each state is free to adopt and adapt CCSSM to best use materials and resources, provided that they can be incorporated within the CCSSM program. As of this writing, forty-five of the fifty states (and three U.S. territories) have adopted these standards and have received funding by governors and state educators to implement the program by around 2014.

Following are some excerpts from a May 2011 letter to the mathematics education community from National Council of Teachers of Mathematics President Michael Shaughnessy. He elaborates on some key features of the standards:

> Systemic changes in mathematics education will necessarily accompany the Common Core State Standards for Mathematics *because these standards are national in character.* The individual states themselves, as is their right under our Constitution, have, one by one, decided to accept the Common Core State Standards. This is new in the United States—never before have so many states agreed to base mathematics instruction on a common set of standards. Furthermore, these standards include both Standards for *Mathematical Content* [emphasis mine] and Standards for *Mathematical Practice* [emphasis mine], and students' mastery of both the content and the practices will be assessed in the designs being created by the two assessment consortia. By the way, the eight Standards for Mathematical Practice are not teaching practices—they are *student practices*—processes that students need to engage in and develop facility with as they learn mathematics and solve mathematical problems. The fact that students will be accountable to the Standards for Mathematical Practice is another systemic change for mathematics education in our country. Although in the past individual states have included performance tasks and extended constructed response items calling for student work and reasoning, such assessments have been inconsistent, and sometimes not persistent, in the United States. This will change.
>
> In light of the national character of CCSSM and the new types of accompanying assessments, we can start doing certain things now as we think about curriculum, instruction, and assessment in the era of the Common Core.

The Common Core lists eight Standards for Mathematical Practice:

1. Make sense of problems and persevere in solving them.
2. Reason abstractly and quantitatively.
3. Construct viable arguments and critique the reasoning of others.
4. Model with mathematics.
5. Use appropriate tools strategically.
6. Attend to precision.
7. Look for and make use of structure.
8. Look for and express regularity in repeated reasoning.

Both this book and CCSSM view learning mathematics as a meaning-making activity rather than as a collection of rules and regulations. For part of that perspective, both focus on student understanding of content rather than mindless repetition (though not all repetition need be mindless). In connection with understanding, problem solving gets high priority and is viewed as a verb because it involves active participation rather than passive acceptance.

In addition to underlying principles of commonality, this book and CCSSM share some content, including the distributive property, the examination of number systems and their extensions (and the relevance of field properties in particular), the interplay of geometry and algebra, the Pythagorean theorem, and prime numbers. This book uses the Pythagorean theorem to offer historical context to support Euclid's implicit appreciation of the child's deep and playful impulse of measurement and uses prime numbers to make the ordinary appear extraordinary.

Although the strategies for solving are helpful, you might also consider important questions of a more personal nature:

- How might this problem have come about?

- If I were to phrase this problem, might I do so differently? How so? Why?

- As I look at a problem not in isolation, but as part of an array, how do I see them as connected?

- What is my emotional reaction in thinking about this particular problem or the array of problems that I have explored or that have been presented for me to explore? Do I find them threatening? Enticing? If I had a choice which one(s) to work on, how would I decide? Would I prefer to work on some alone, or with other people? What qualities of mind/compassion/attitude would I choose in people with whom I would wish to collaborate?

- How much time should I (be expected to) work on this problem or an array of problems? Is trying it for an hour or two worthwhile, or should I devote a few days? A lifetime? (Is the last question absurd, or merely an indication of what we do not feature in most educational settings?)

- What sort of aesthetic appeal do I find in the variety of problems and their solutions that I have worked on? Is aesthetic appeal important?

Not only do these bulleted questions open the terrain for perceiving problems in a humanistic light, they also suggest that we can view problems as something more than an invitation for solution. What other options exist? We could use a problem to pose other problems, unconnected with the desire to solve a given problem. Or we could "neutralize" problems so that what was a problem could become a situation not focused on solutions. Something that is neutralized does not lend itself to answering questions such as "is it true or false?" or "does it logically hold?" Such questions do not usually make sense regarding situations. It is one thing, for example, to be presented with the problem of showing that the square on the hypotenuse of a right triangle is equal to the sum of the squares on the other two sides. It is another to reframe this problem as a situation—for example, that we have a right triangle, or even that we have a right triangle with squares erected on the three sides. It is yet another act of neutralizing the Pythagorean theorem to observe that we have a right triangle that has some objects erected on the three sides. This brings us back to chapter 7: "noticing" as an important educational achievement.

"De-probleming" the problem into a situation can tell us what we consider to be mathematically salient. Analyzing how different people pose and neutralize problems can offer a better understanding of diverse opinions and even worldviews. Thinking about some of the above questions not only can lead to a better understanding of specific problems and topics but also may shed light on idiosyncratic ways in which we each construe mathematics as a subject, and how we connect it with other fields and life experiences.

Problem posing also can encourage us to assume a modified attitude toward knowledge in general. Knowledge is not only what we are given by others, especially experts. It is also something that even the neophyte can hope to acquire. It can be achieved not only with esoteric subject matters but also by looking at familiar territory in new ways through a "what if not" perspective.

Statement 6 from the CCSSM Standards for Mathematical Practice expresses a recurring theme on the quality of mathematical thinking:

> **6.** Attend to precision. Mathematically proficient students try to communicate precisely. They try to use clear definitions in discussion with others and in their own reasoning. They state the meaning of the symbols they choose, including using the equal sign consistently and appropriately. They are careful about specifying units of measure, and labeling axes to clarify the correspondence with quantities in a problem. They calculate accurately and efficiently, express numerical answers with a degree of precision appropriate for the problem context. In the elementary grades, students give carefully formulated explanations to each other. By the time they reach high school they have learned to examine claims and make explicit use of definitions.

Encouraging precision and its close relative clarity are important goals. But how should we think about them? When is it appropriate to lessen their grip in exploring mathematical ideas? Lest we overstate the role of precision in CCSSM's agenda, the description in statement 6 does indicate a healthy sensitivity to context with regard to precision. It speaks about "degree of precision appropriate for the problem context." That realization opens up large terrain to explore.

The following will present ways of thinking that do not necessarily deny or devalue the importance of precision and clarity, but rather suggest how they might be tempered to unleash emotional and personal concomitants of an otherwise predominantly cognitive focus.

Expanding the Emotional and Personal Intertwine

We will open up the terrain to include important findings of a personal nature that incorporate contributions at the forefront of knowledge. When we focus too heavily on clarity, do we not overlook that almost all beginnings (and beginnings appear regularly as we continue to assimilate new ideas) lack clarity and are possessed with doubt and uncertainty? The less trivial the subject matter, the more such qualities pervade our psyche.

Walker Percy, speaking with *New York Times* correspondent Kim Heron (1987) about his novel *The Thanatos Syndrome*, attempts to locate the cause of much anxiety and depression in society: "A great deal of the anxiety and depression people experience comes from watching television, where there's a certain aesthetic unity—everything works out. My life and yours are much more fragmentary, haphazard and incomplete" (p. 22). Might this insight suggest a less stringent role for clarity as we engage in new territory? Actually, if we have a broader mindset, does not all territory have the potential to be new if we learn to pose problems more often? Even if we approach problem solving and its cohorts in less dramatic psychological terms, do we not overplay the role of clarity as a fundamental organizing element in our thinking?

Referring to reflective thinking as a near relative of problem solving, John Dewey (1933), perhaps one of the most outstanding educational philosophers of the twentieth century, comments: "Reflective thinking . . . involves (1) a state of doubt, hesitation, perplexity, mental difficulty, in which thinking originates, and (2) an act of searching, hunting, inquiring, to find material that will resolve the doubt, settle and dispose of the perplexity. . . . Thinking begins in what may fairly enough be called a forked-road situation that is ambiguous, that presents a dilemma, that proposes alternatives. As long as our activity glides smoothly along from one thing to another . . . there is no call for reflection" (pp. 12–14).

This book presents many illustrations that indicate the need to live with, rather than dismiss, ambiguity and perplexity. We are all occasionally enticed into seeing something that may be confusing or that is not well developed early on. Rather than being a turn-off, confusion may invite further exploration, such as when we confront a work of art that we can make little sense of at the moment but are intrigued enough to return to over and over, seeing more each time. Confusion, imprecision, and lack of clarity can be powerful motivational forces that tug at the heart of rationality.

Furthermore, these qualities are unavoidable, because no matter how clearly teachers and texts strive to inculcate clarity, every person creates and perceives the world in idiosyncratic ways—ways that may have great value in enabling one to cope with a barrage of information and new worldviews.

Personal Perspective

In college I took a course on finite dimensional vector spaces from a renowned mathematician. The teacher listed a set of axioms early in the course and later got stuck in using them as he struggled over an algebraic proof of a theorem. He reminded the class that it was necessary to make proofs and theorems dependent on purely abstract formulations—axioms, undefined terms, previously proved theorems, and logical deduction. His back was to the class, and I was sitting close enough to the edge of the first row to see how he handled his confusion. He drew a few quick sketches that had no algebraic formulation and then quickly erased them to avoid disclosing the "weakness" of having to call upon nonalgebraic representations. After erasing all his sketches, he went back to his formal proof and reminded us that he had just proven a theorem that depended solely on "axioms, previous theorems, and careful logic."

He seemed embarrassed to use representations on a different plane from the more abstract one he wanted to convey. He was holding on to a scheme that was extremely helpful as a heuristic but would not share it with his students. He needed something at the moment that was at odds with the deity of abstraction that he was presenting to the class explicitly.

We could all profit from discovering and disclosing the ways we operate that help us in holding and generating ideas that may be vague, or at odds with a more precise formulation of the situation or the problem.

An appreciation for less stringent depictions of mathematics, at least as new ideas unfold, is helpful in searching for an intuitive grasp of complex ideas that are accessible in more primitive form before they are assimilated more precisely. The very formulation of the distributive property is a case in point. As chapter 2 showed, people have an implicit understanding of the concept even before it is clearly formulated as a property of numbers.

Historical Perspective

The very names of number systems (natural, negative, real, imaginary, complex) testify that new and extended systems were problematic early in their development. Our formal education often overlooks that the gestation period for developing new ideas is more fragile than is acknowledged. Extensions of number systems often commit the "nominalistic fallacy"—assuming that by giving something a name (such as "imaginary numbers"), it is brought into existence—even if we had been convinced beforehand that it could not exist.

So, the impossible becomes possible by bestowing a label. Some historical perspective indicates that the going may be much rougher than is usually conveyed. Making a leap that had been previously rejected for a variety of reasons sometimes takes years. János Bolyai (1802–1860) was a central figure in creating an alternative to Euclidean geometry. Non-Euclidean geometry required taking what was self-evident (through a point not on a line, there must exist exactly one other line parallel to the given one) and standing it on its head. Bolyai's father, himself a mathematician, wrote the following to his son, who was plagued by doubts in developing non-Euclidean geometry: "For God's sake, I beseech you, give it up. Fear it no less than sensual passions because it, too, may take all your time, and deprive you of your health, peace of mind, and happiness in life" (Boyer 1968, p. 589). This language is strong and even funny until we realize that these comments were made before the sexual revolution of the last quarter of the twentieth century. The point is that many mathematical extensions required not only new insight but also great courage. People did not want to relinquish the simpler systems; at stake was the notion that desired extensions often ran counter to what was accepted to be logically/ intuitively true. Many viewed adopting new perspectives as being on shaky ground and even pure folly. Finding out that creating new ways of looking at the world is often fraught with pain and threat should console those of us who are perplexed as we take on new ways of seeing and creating.

Consider another example of a more recent discovery that perplexed and upset the mathematical community for more than a century. It is about a problem introduced in 1904 and, despite tremendous effort to solve it, was not resolved until 2006. The story appears in *Poincaré's Prize*, by George Szpiro (2007). In 1904, a French mathematician, Jules Henri Poincaré, proposed a conjecture about the behavior of rubberized spheres and distorted spheres (ones that could be morphed into spheres by twisting and stretching, but not by tearing and gluing) in different dimensions—a topology problem. A bagel and a sphere, then, would not be topologically equivalent. To appreciate an accurate, humorous rendition of the field, Szpiro offers the following: "Three mathematicians are shown a cube and asked to describe what they see. The first, a geometer, says, 'I see a cube.' The second is a graph theorist. She ventures, 'I see eight points connected by twelve edges.' The third, a topologist, declares, 'I see a sphere'" (p. 53). Szpiro summarizes a popular version of the topology problem: "Imagine an ant crawling around on a large surface. How would it know whether the surface is a flat surface, a round sphere, or a bagel-shaped object? The ant would need to lift off from the surface to observe the object from afar, so how could one prove the shape was spherical without actually seeing it? Raise the surface to the next higher dimension, and you have the problem that Poincaré sought to solve" (inside front cover).

Many texts include popular renditions of topology as well as the statement of the problem (see Ornes 2006 for an accessible description). Though partial headway was made in proving the conjecture, an uncontested proof did not emerge for a long time. German-born Russian mathematician Grigory Perelman came up with one a century later, in 2004. This story has much drama. Perelman was awarded the Fields Medal—the equivalent of a Nobel Prize in science—to be presented in 2006 in Madrid, Spain. Perelman was living in relative poverty with his mother in St. Petersburg. Unaffiliated with an academic setting, he neglected to appear for the award. More astounding, however, he refused the award's $1 million grant. He felt isolated from the mathematical community and felt that many members of the community had stretched ethical bounds beyond what should be tolerated.

Szpiro (2007) comments, "According to a former colleague . . . who prefers to remain anonymous, Perelman is so deeply disappointed by the perceived decline of ethics in the mathematics community that he no longer considers himself a professional mathematician. . . . But true to his character, he abhors controversies, and rather than get involved in disputes that seemed inevitable he apparently preferred to cut all ties to his former colleagues" (pp. 6–7). Though the circumstances underlying Perelman's attitude are now shrouded in mystery, uncovering them later might be valuable.

The history of efforts to prove the conjecture is filled with much poignancy. One involves my former mentor and colleague Edwin E. Moïse. Below are excerpts from Szpiro's (2007) book.

> Moïse served in the US navy in World War II; he was a member of the group that broke the Japanese naval code. . . . In the early 1950s, after completing his Ph.D. [under R.L. Moore], Moïse was invited to the Institute for Advanced Study [at Princeton]. . . . He said he was going to prove Poincaré's Conjecture. And he did not. Both Moïse and Papa soon became aware that the other was working on the same problem. The competition . . . was intense. One day Papa announced—to Moïse's consternation—that he had solved the problem. . . . In due course a hole was found . . . and Moïse could breathe a sigh of relief. A few weeks later Moïse announced that he had found a solution. Of course, now a hole was found in his "proof" and it was Papa's turn to breath a sigh.
>
> In short, Moïse made no headway though he spent a number of years in the pursuit. One of his students announced that eventually he seemed resigned that he would not be successful.
>
> He never came to grips with his defeat. When it finally dawned upon him that he was not going to be the one to crack the conjecture, he turned away from mathematical research altogether.
>
> Moïse moved to literary criticism. He published six short notes on nineteenth century poets. . . . After suffering two strokes, Moïse died in December 1998. (pp. 132–34)

Of course, not all efforts at mathematical exploration were as heavily laden with the frustration described by Bolyai for non-Euclidean geometry, Perelman's sense of isolation, and Moïse's efforts with the Poincaré conjecture. Others were uplifting—perhaps more so for the discoverers than their oxen. Though it may be apocryphal, a well-repeated tale in mathematical circles claims that when the Pythagorean theorem was discovered, more than two millennia ago, Pythagoras and his followers were so delighted that they sacrificed one hundred oxen at the altar out of joy. Rumor has it that now whenever a great mathematical discovery is made, all the oxen in the world shudder.

When incorporating new ways of thinking about mathematical ideas, people might be comforted to know or at least imagine what difficulties those who first tried to make sense of these new approaches confronted. Can we imagine what mathematical circumstances might have precipitated strong emotional (joy, pain, fear) reactions when they attempted to extend systems beyond those already well accepted? We might refer to the activity of hypothesizing about such circumstances (even without being presented with information about the actual confusions and controversies at the time) as pseudohistory.

This activity is one that students working together and trying to share their joys and frustrations can carry out. We all could learn to become aware of one another's idiosyncratic ways of bringing new ideas into existence and to share and respect the labor pains that accompany the process.

A Gloss on Style of Exposition

We need not adopt an either–or perspective for rationality and emotionality, clarity and confusion, and personal and public ways of thinking. How can they be creatively integrated within the minds of both T. C. Pits and professional educators?

The philosopher Israel Scheffler (1993) puts dreaming/daydreaming and "what if not" thinking more generally in a perspective that gives it loftier educational coinage than is generally acknowledged. He says:

> Students of thinking tell us . . . that . . . dreams may do intellectual work, and the history of science provides ample evidence. A striking case . . . is Kekulé's discovery of the ring structure of the benzene molecule, after dozing in front of the fire and dreaming of snakes dancing in ringlike fashion. The notion of the incubation of a problem, where without conscious attention to it, the mind silently works out a solution, has been long credited in experience. Sometimes the best approach to a problem is to turn away from it completely, let the mental machinery idle, go for a walk, take in a movie, have a cup of cocoa. (p. 115)

Scheffler considers more than dreaming in locating traits that are educationally valuable, even though they are often thought of as vices (such as forgetting, procrastination, pointlessness). Distinguishing between knowledge as a collective heritage and the acquisition of knowledge, he says:

Knowledge as a collective heritage of recorded information is indeed a fundamental resource of the teacher, but it cannot be transferred bit by bit in growing accumulation within the student's mind. The teacher must strive rather to promote an insight into the meaning, basis and use of the collective heritage, so that the student may in fact come to know it rather than simply being informed of it. Even such knowing is not enough to express the aims of education, however, for it leaves out the opportunity for *innovation by the learner*, [the] ability to go beyond a knowing of available truths. We do not feed into the learner's mind all that we hope of available truths . . . as an end result of our teaching. . . . Our pupils can and should be expected to gain new understandings beyond our present grasp; they will need to revise our science, expand and modify our scholarship, recast our social, historical, and legal suppositions. They will, in short, need to discover new trends beyond our ken. (p. 107)

A Limit to Rationality

I have been suggesting that it might be worth incorporating within the educational scene elements of thinking, behaving, and feeling that are normally associated with the humanities: metaphor, humor, dreams, storytelling, personal reflection. Many people believe that mathematics does not share an important quality of the humanities and the social sciences: that some problems will never be solved. Some examples: What is a good person? What sort of life should we live? What was the relationship among the first few bipeds and how do their ways of communicating compare with the modern evolution of the species? How will the most salient elements of a democratic society evolve? How much freedom should I allow in the bringing up of my children? How much are we as a nation on a course to self-annihilation?

A popular belief maintains that all mathematically well-formulated problems can be solved. It may take time (as suggested by the history of the Poincaré conjecture); it may take the mind of a genius to solve; it may have to be reformulated to lend itself to a solution. It may drive investigators crazy in their attempts at solution, but many people believe that, in theory at least, no unsolvable problems exist.

That point of view has been challenged and the answer is surprising—one whose implications have not been adequately explored. In the first third of the twentieth century, an Austrian philosopher of mathematics, Kurt Gödel (1906–1978), took on the challenge of a German mathematician, David Hilbert (1862–1943). Hilbert wanted to establish a firm foundation for mathematics, which involved two forks: (1) that it was possible to prove (with airtight rigor) that mathematics had no inconsistencies (and that any supposed inconsistencies had simply not been adequately worked through) and (2) that all mathematical problems were theoretically solvable.

Gödel showed not only that the longing for a demonstration of absolute consistency (in a system as ostensibly mild as the set of natural numbers) is a pipe dream but also that the prospect of proving or disproving many statements within the system is impossible. Can you imagine some all-knowing being bestowing such a tantalizing treasure before us . . . chuckling away with the whisper saying, "And I ain't gonna tell when you spend a lifetime trying unsuccessfully to prove something and come upon a stone wall whether the essential difficulty is that you have not worked hard enough or that the problem is fundamentally unsolvable!" That is, "undecidable" statements must exist in any such system—statements that are true but unprovable as such.

How to interpret his findings and exactly what they might mean is a problem from both an educational and a mathematical point of view. Clearly, however, Gödel has in some sense hoisted rigor and hung it on its own petard.

For further elaboration on Gödel's work and its implications in other areas, see Nagel and Newman (1958), Hofstadter (1979), and Dawson (1999).

We are drawn then not to a proposal or resolution, but to a deep finding that may make disentangling mathematics and the humanities at their deepest levels impossible.

How much of the content suggested in this book can be incorporated within an already chock-full curriculum is a matter of debate and experimentation. Some of the spirit of inquiry, however, may be adopted within the bounds of present and emerging revisions of curriculum. An appreciation for and acknowledgment of some of the personal, emotional, and humanistic qualities this book discusses should broaden what is meant by experiencing mathematics. Questions of the following sort have emerged from this perspective:

- How much leeway (including misconceptions and deviations from the "correct" interpretation) should we allow and encourage at various stages of learning new content or of rethinking old?

- What difficulties were experienced by mathematicians who tried to tackle some of these ideas originally—especially ones passed along today as noncontroversial (as in our discussion of emerging number systems and in the development of geometry within a limited view of number systems)?

- How might we imagine the difficulties experienced even if we do not know the actual history (hypothesizing what the historical issues might have been)?

- What images and metaphors do people grasp as they try to make sense of and learn new content?

- How might humor function in mathematical inquiry (as discussed in our view of many of the conjectures and theorems in number theory)?

- What brief stories can people devise that capture the mathematical ideas in a nontechnical way?

- How would you describe the many ways this book used the distributive property? Some sections described what the distributive property itself was. Some explored the role of the property as embedded in other systems. Some explored how the property offered a sharper focus on a puzzling situation. Sometimes the property was used to illustrate the connection between different branches of mathematics.

What topic(s) other than the distributive property might be used to illuminate diverse aspects of your mathematical experience? I look forward to reading about your suggestions.

References

Boyer, Carl B. *History of Mathematics.* Boston: John Wiley and Sons, 1968.

Dawson, John W. "Gödel and the Limits of Logic." *Scientific American* 280, no. 6 (1999): 76–81.

Dewey, John. *How We Think.* Boston: DC Heath and Company, 1933.

Heron, Kim. "Technological Hubris." *New York Times Book Review* (April 5, 1987): p. 22.

Hofstadter, Douglas. *Gödel, Escher, and Bach: An Eternal Golden Braid.* New York: Vintage Books, 1979.

Lieber, Lillian. *The Education of T. C. Mits.* New York: W. W. Norton, 1942.

Nagel, James, and James Newman. *Gödel's Proof.* New York: New York University Press, 1958.

National Governors Association Center for Best Practices (NGA Center) and Council of Chief State School Officers (CCSSO). *Common Core State Standards for Mathematics. Common Core State Standards (College- and Career-Readiness Standards and K–12 Standards in English Language Arts and Math).* Washington, D.C.: NGA Center and CCSSO, 2010. http://www.corestandards.org/.

Ornes, Stephen. "What Is the Poincaré Conjecture?" 2006. http://seedmagazine.com/content/article/what_is_the_poincare_conjecture/.

Scheffler, Israel. "Vice into Virtue, or Seven Deadly Sins of Education Redeemed." In *Problem Posing: Reflections and Applications*, edited by Stephen I. Brown and Marion I. Walter, pp. 104–16. Mahwah, N.J.: Lawrence Erlbaum Associates, 1993.

Shaughnessy, Michael. *Summing Up.* Reston, Va.: National Council of Teachers of Mathematics, May 2011.

Szpiro, George. *Poincaré's Prize.* New York: Dutton Publishers, 2007.